BRITANNICA

Mathematics in Context

Operations

Britannica
ENCYCLOPÆDIA BRITANNICA EDUCATIONAL CORPORATION

Mathematics in Context is a comprehensive curriculum for the middle grades. It was developed in collaboration with the Wisconsin Center for Education Research, School of Education, University of Wisconsin–Madison and the Freudenthal Institute at the University of Utrecht, The Netherlands, with the support of National Science Foundation Grant No. 9054928.

National Science Foundation

Opinions expressed are those of the authors
and not necessarily those of the Foundation

ISBN 0-7826-1524-4
1 2 3 4 5 WK 02 01 00 99 98

The *Mathematics in Context* Development Team

Mathematics in Context is a comprehensive curriculum for the middle grades. The National Science Foundation funded the National Center for Research in Mathematical Sciences Education at the University of Wisconsin–Madison to develop and field-test the materials from 1991 through 1996. The Freudenthal Institute at the University of Utrecht in The Netherlands, as a subcontractor, collaborated with the University of Wisconsin–Madison on the development of the curriculum.

The initial version of *Operations* was developed by Mieke Abels and Monica Wijers. It was adapted for use in American schools by Gail Burrill, Aaron N. Simon, and Beth R. Cole.

National Center for Research in Mathematical Sciences Education Staff

Thomas A. Romberg
Director

Joan Daniels Pedro
Assistant to the Director

Gail Burrill
Coordinator
Field Test Materials

Margaret R. Meyer
Coordinator
Pilot Test Materials

Mary Ann Fix
Editorial Coordinator

Sherian Foster
Editorial Coordinator

James A. Middleton
Pilot Test Coordinator

Project Staff

Jonathan Brendefur
Laura J. Brinker
James Browne
Jack Burrill
Rose Byrd
Peter Christiansen
Barbara Clarke
Doug Clarke
Beth R. Cole
Fae Dremock
Jasmina Milinkovic
Margaret A. Pligge
Mary C. Shafer
Julia A. Shew
Aaron N. Simon
Marvin Smith
Stephanie Z. Smith
Mary S. Spence

Freudenthal Institute Staff

Jan de Lange
Director

Els Feijs
Coordinator

Martin van Reeuwijk
Coordinator

Project Staff

Mieke Abels
Nina Boswinkel
Frans van Galen
Koeno Gravemeijer
Marja van den Heuvel-Panhuizen
Jan Auke de Jong
Vincent Jonker
Ronald Keijzer
Martin Kindt
Jansie Niehaus
Nanda Querelle
Anton Roodhardt
Leen Streefland
Adri Treffers
Monica Wijers
Astrid de Wild

Table of Contents

Dear Student,

Have you ever used positive and negative numbers?

In this unit, you will use positive and negative numbers to draw designs like ones made in India.

You will use a world map to explore time zones and figure out the best times to call people in other parts of the world.

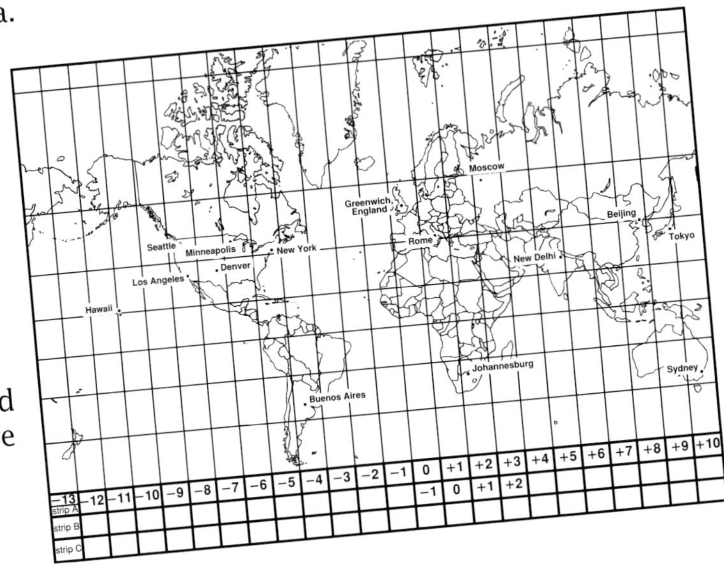

You will also play a game that will help you learn how to add and subtract positive and negative numbers.

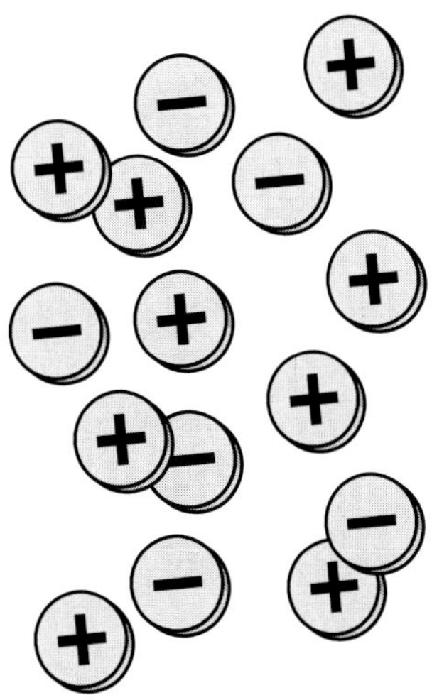

You will multiply and divide positive and negative numbers in order to find average temperatures.

In the last section, you will investigate how to move, enlarge, and reduce a shape on graph paper using positive and negative numbers.

We hope you enjoy this unit and learn a lot about operations with positive and negative numbers.

Sincerely,

The Mathematics in Context Development Team

A. GRAPHING FIGURES

Veda Designs

Some Middle Eastern art is based on mathematics from the *Vedas*, which are books from ancient India.

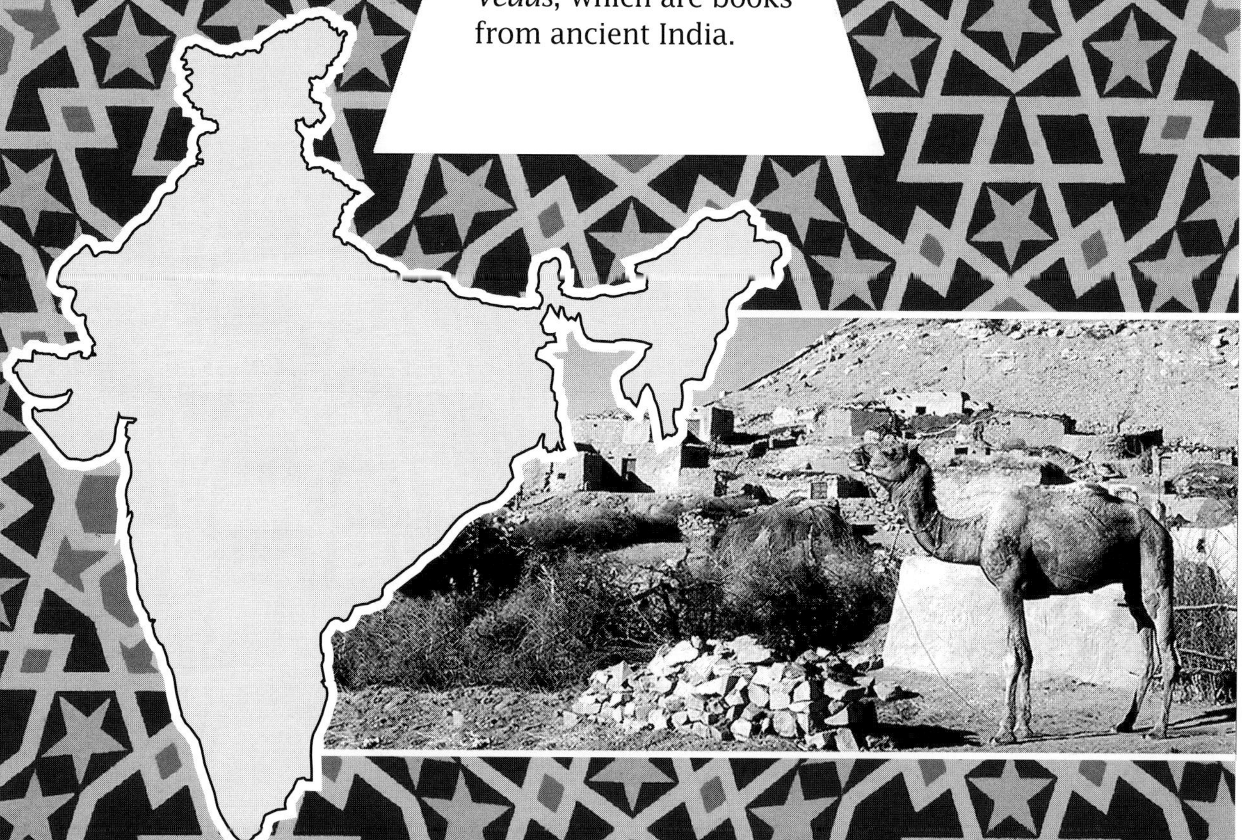

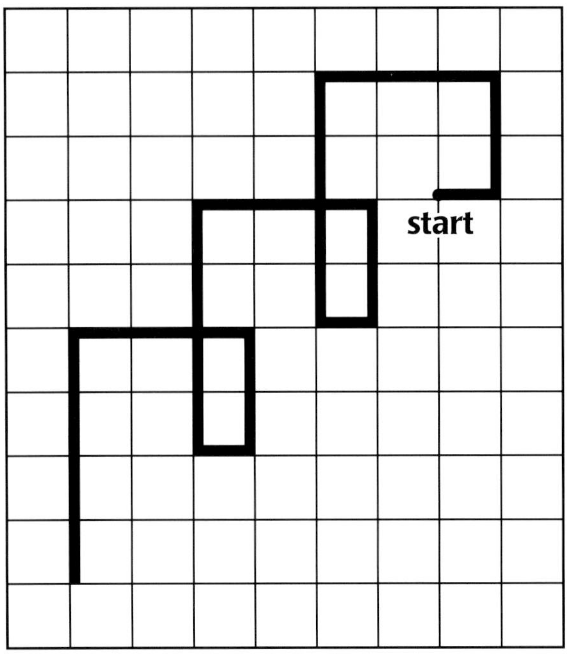

The design on the left is similar to some of the Veda designs. It "walks" down and to the left.

1. Describe how the design on the left was drawn.

2. Below you see another design. Draw this design on the grid paper on **Student Activity Sheet 1** and extend the pattern. How does this design walk?

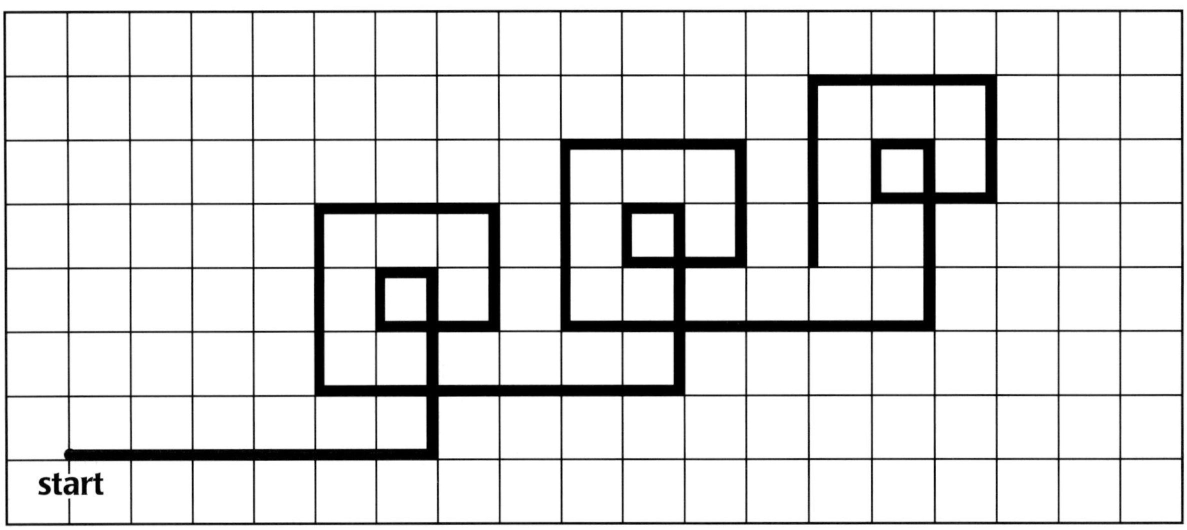

One way to describe designs is with arrows.

→1 ↑2 ←3 ↓2 →1 ↑1 ←2 ↓3 →2 ↑1

←1 ↓2 →3 ↑2 ←1 ↓1 →2 ↑3 ←2 ↓1

3. a. What could the arrows mean?

b. What could the numbers mean?

c. Draw Design 1, indicated by the above numbers, on **Student Activity Sheet 1.**

4. a. Now draw Design 2, indicated by the numbers below. Does this design walk in a particular direction?

→5 ↑4 ←2 ↓5 →4 ↑2 ←5 ↓4 →2 ↑5 ←4 ↓2

b. Is there a pattern in the numbers?

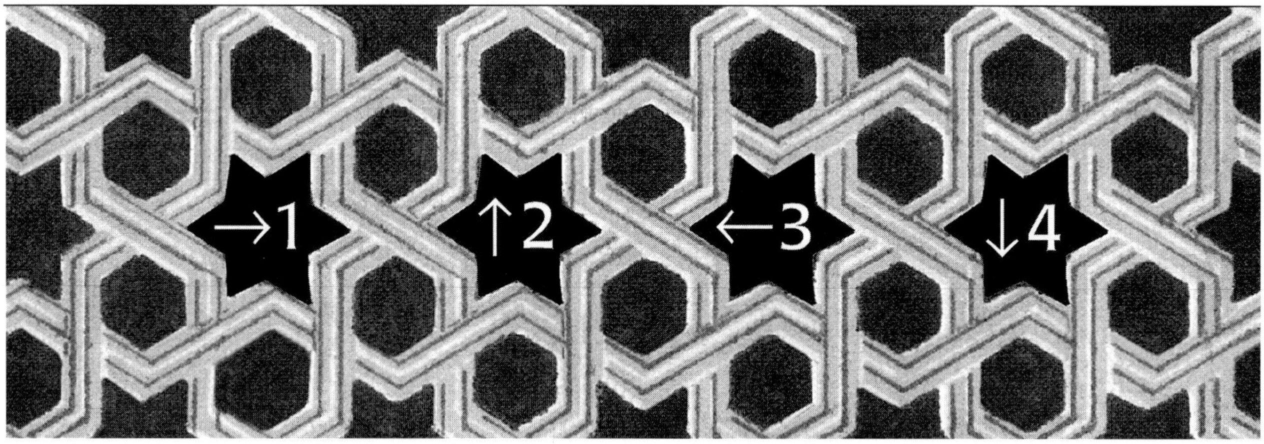

You may have noticed that the *directions* of the steps are always in the same order. From the starting point, you move right, up, left, down, right, up, left, down, and so on. All of the designs are drawn this way, so you don't have to draw the arrows every time.

Suppose that the drawing instructions for a design started with the numbers below.

1 1 4 1 1 4 … … … …

5. Draw the beginning of this design. Finish the design so that you end where you started.

6. a. Create your own design on graph paper.

b. Write drawing instructions for your design.

c. Is there a pattern in your numbers? If so, what is it?

7. a. What is the pattern in the row of numbers below?

1 1 3 3 1 1 3 3 … … …

b. What would the design look like if you extended it by six more numbers?

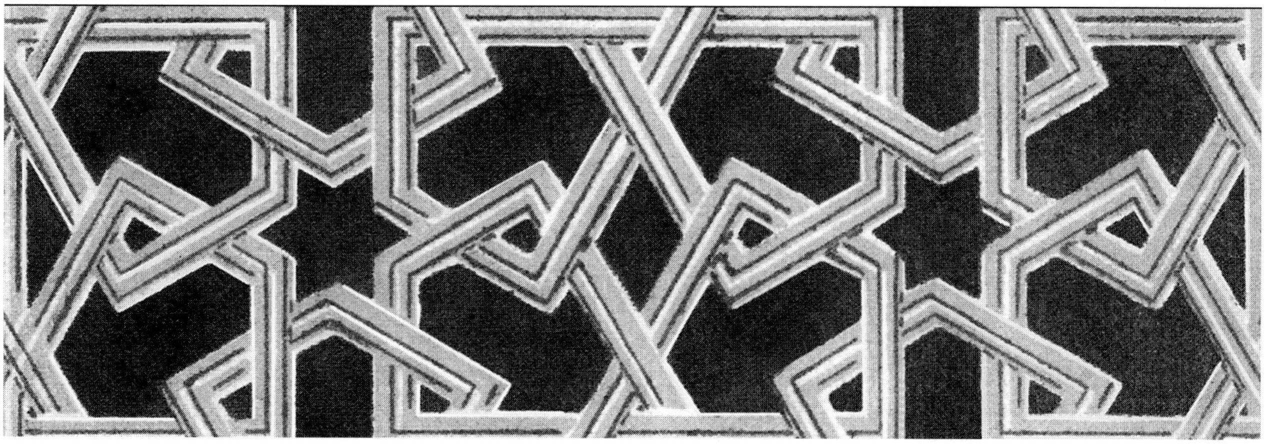

8. a. What is the pattern in the row of numbers below?

2   3   4   4   3   2   ...   .

b. What might the next six numbers be? Draw this design.

c. Are there any other possibilities for the next six numbers that will produce a design that ends at the starting point?

9. a. On the graph below you see four points of a design. The design starts at *O*, goes from *O* to *A* in four moves, goes from *A* to *B* in the same four moves, and so on. What might the drawing instructions be for this design?

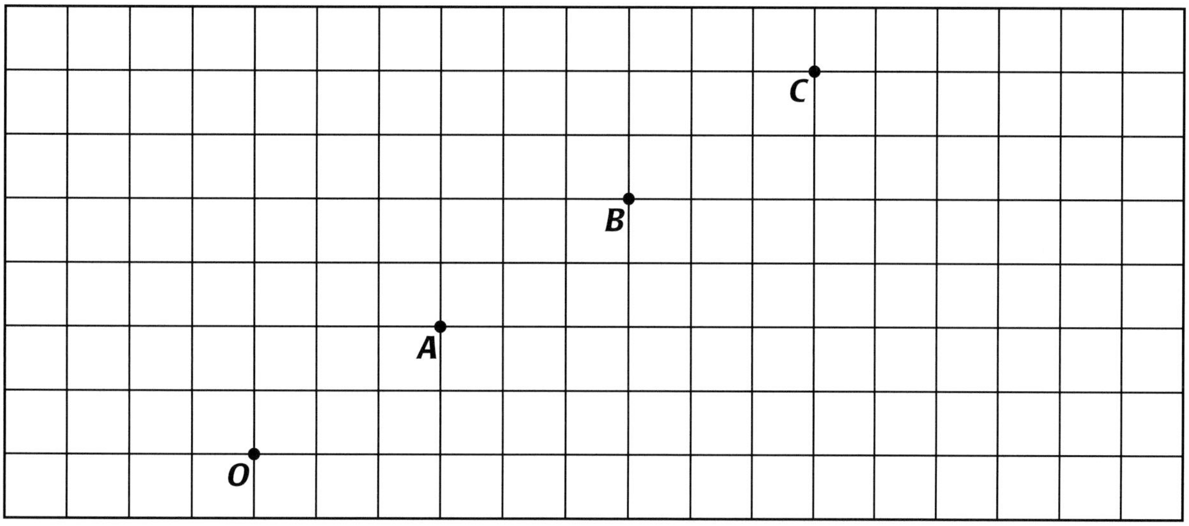

b. Do you think everyone in your class wrote the same drawing instructions? Explain.

After you drew a few designs, you probably noticed that some designs end where they start and others continue forever (they "walk" off the paper). Because some designs walk off the paper, you need to be careful where you start them.

Daisy made up some design instructions.

1 3 2 1 1 3 2 1 1 3 2 1

Daisy wants to know where to start her drawing so that she'll be able to see if it walks off the paper. To see if the design walks to the right or the left, Daisy circles some of the numbers.

① 3 ② 1 ① 3 ② 1 ① 3 ② 1

10. **a.** How do the circled numbers relate to the directions?

b. How can Daisy tell whether the design walks to the right or the left without drawing it?

c. How can you tell if the design moves up or down without drawing it?

d. Where on the paper should Daisy start drawing her design?

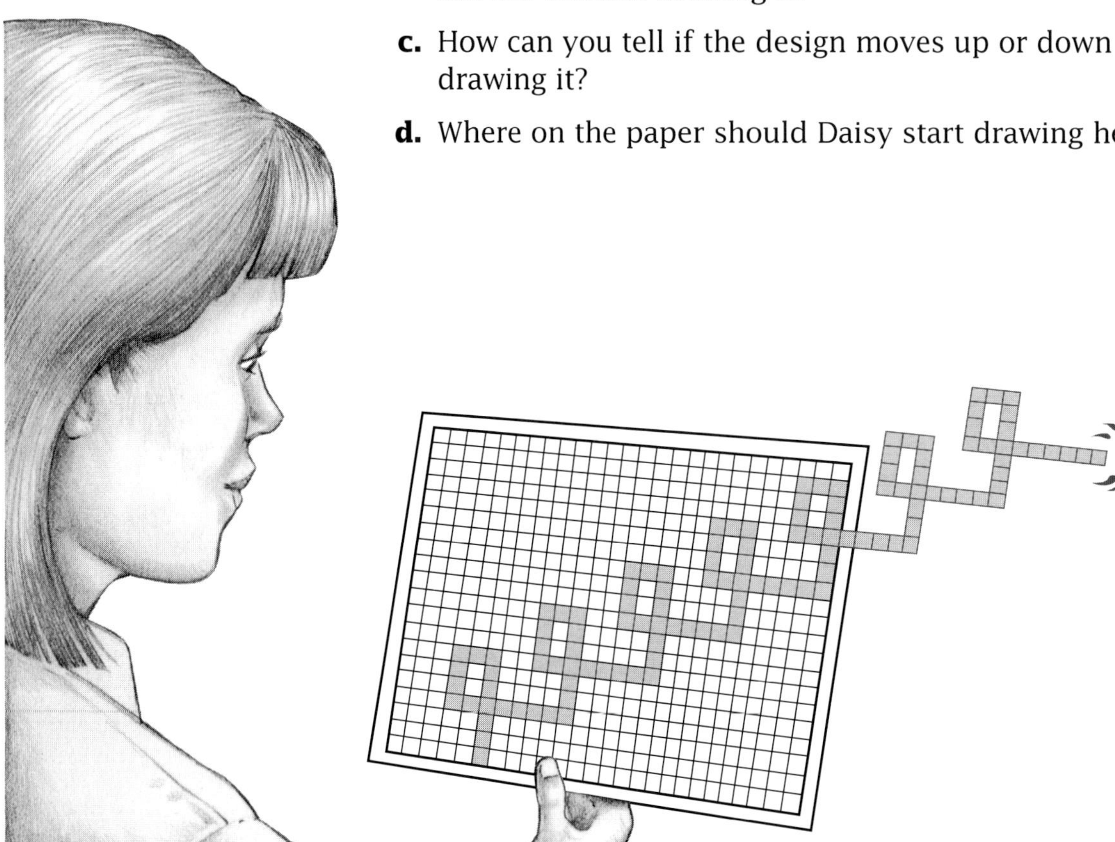

Below are the instructions for two designs.

Design 1:	4	3	3	2	2	3	3	4
Design 2:	4	3	3	4	4	3	3	4

11. a. One of the designs ends where it starts. Can you tell which one it is without making a drawing? Explain how you figured it out.

b. Explain how you can tell in what directions the other design walks without drawing it.

Pluses and minuses can replace the arrows as another way of writing instructions for a design.

Design 3: +4 +1 −1 −3 +4 +1 −1 −3 ...

12. a. What do the pluses and minuses tell you about the directions?

b. Draw Design 3.

Look at the design instructions below.

Design 4: +1 +3 −3 −2 +1 +3 −3 −2 ...

13. Predict in what directions the design walks without drawing it.

Here are three design instructions.

Design A:	+4	+2	−1	−1	+4	+2	−1	−1	You are now at point *A*.
Design B:	+1	+1	−4	−2	+1	+1	−4	−2	You are now at point *B*.
Design C:	+1	+3	−2	−1	+1	+3	−2	−1	You are now at point *C*.

14. a. Draw a point in the middle of a sheet of grid paper. This will be your starting point, *O*. Now draw the three designs above, each starting at *O*. Label the endpoint of each design with its corresponding letter.

b. Where is point *A* in relation to where you started? Where are points *B* and *C* in relation to where you started?

15. Write drawing instructions for a design that walks in the direction not taken by the three designs above. Describe the location of its endpoint, *D*, with respect to point *O*.

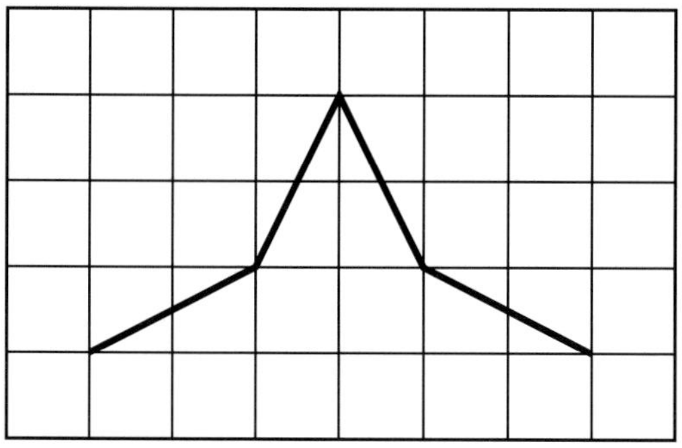

The designs you have studied so far moved only in horizontal and vertical directions. Suppose, instead, that you wanted to make drawings with slanted lines. What drawing instructions could you use? Look, for example, at the drawing on the left.

16. How could you invent instructions for this drawing?

Make up a drawing that includes slanted lines. On a separate piece of paper write the drawing instructions. Give the instructions to a classmate and ask him or her to make the drawing using only your instructions.

17. Did your instructions work? Why or why not?

Summary

In this section, you drew designs on grid paper and used different notations for directions. A "+" meant a move to the right or up, while a "−" meant a move to the left or down. Numbers indicated the lengths of each move.

At the end of the section, you wrote instructions for drawing slanted lines.

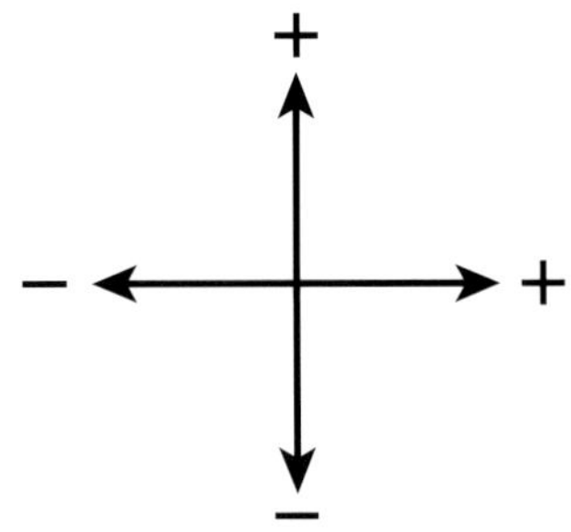

Summary Questions

18. **a.** How can you tell that the following design will return to its starting point?

+1 +2 −3 −1 +2 +3 −1 −2 +3 +1 −2 −3

b. Draw the design.

19. Describe the different ways you represented designs with numbers. Which do you like the best?

20. **a.** Create a design that walks up and to the right.

b. Write drawing instructions for your design using plus and minus signs with numbers.

c. How can you tell just from the drawing instructions that your design walks up and to the right?

B. TIME ZONES

What Time Is It There?

1 Harold and Felicia are big tennis fans. They are watching the Australian Open final being transmitted directly from Australia. Suddenly, their little brother enters the room.

2 It is 7:30 P.M., and Peter knows that tomorrow his cousin Susan is giving her first solo piano recital. He calls her in London to wish her luck.

MARY IS FLYING FROM MINNEAPOLIS TO SEATTLE. HERE IS HER FLIGHT INFORMATION.

Flight	Date	From/To	Time
NW1607	12/05	Minneapolis to Seattle	11:30 A.M. to 1:00 P.M.

MARY IS HAPPY THAT HER TRIP ONLY TAKES $1\frac{1}{2}$ HOURS.

THEN SHE READS THE SCHEDULE FOR HER RETURN FLIGHT.

Flight	Date	From/To	Time
NW0008	12/12	Seattle to Minneapolis	11:45 A.M. to 5:15 P.M.

MARY WONDERS WHY THE TRIP BACK FROM SEATTLE TAKES SO MUCH LONGER THAN THE TRIP TO SEATTLE.

1. These three stories have something in common. Can you explain what it is? This photo of Earth might help.
2. Have you ever had an experience like those in the three stories? If so, describe it.

Courtesy of National Aeronautics and Space Administration.

Activity

WORLD TIME ZONES

3. Use **Student Activity Sheet 2** to make the cylinder and the ring you see below. The cylinder is used as a three-dimensional model of Earth that can help you answer problem **4** below.

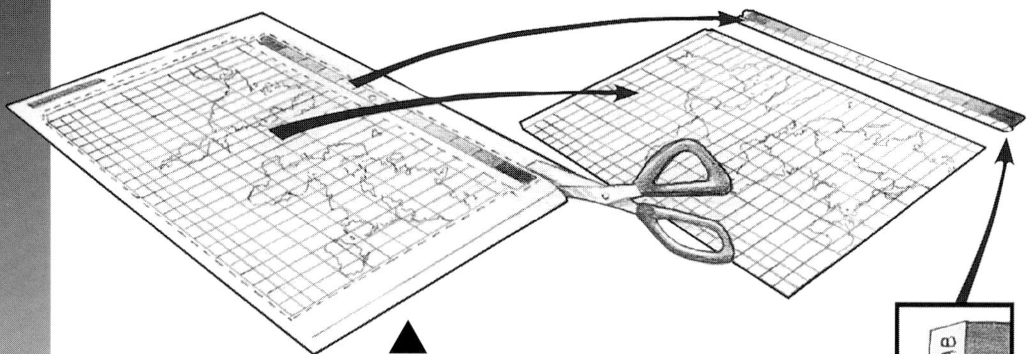

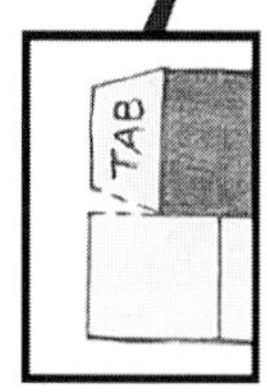

i. Cut out the map and the sun ring on the student activity sheet. Do not cut off the tabs!

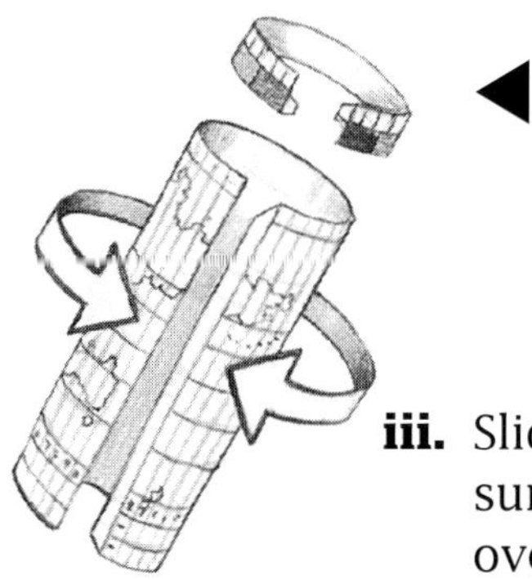

ii. Roll the map into a cylinder and tape it closed over the tab. Do the same for the sun ring.

iii. Slide the sun ring over the cylinder.

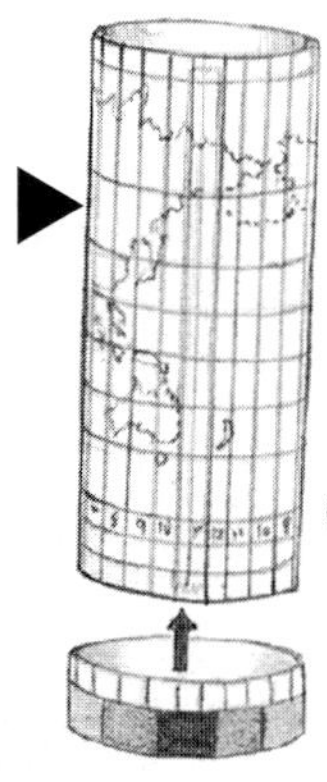

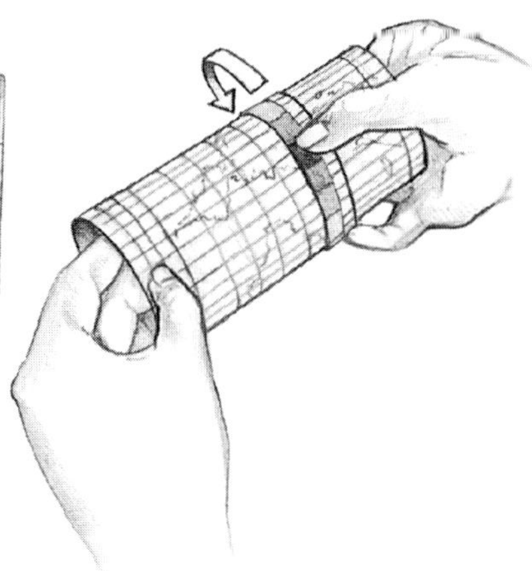

4. **a.** How can the sun ring on your model explain the photo on the previous page?

b. How can you use the sun ring on your model to explain the first and second stories on page 11?

5. **a.** When it is noon in New York, where is it midnight?

   b. What is the time difference between New York and Los Angeles?

   c. What is the time difference between London and New York?

Below is a world map showing the international time zones. You can find a map like this in an atlas.

> Time is calculated from a *zero line* in Greenwich, England. Each zone is 15° across. When you move from one time zone to the next, you have to change your watch by one hour either backwards or forwards, depending on which way you are traveling.

At the bottom of the map is a strip that shows how the time zones are related to the zero-time zone.

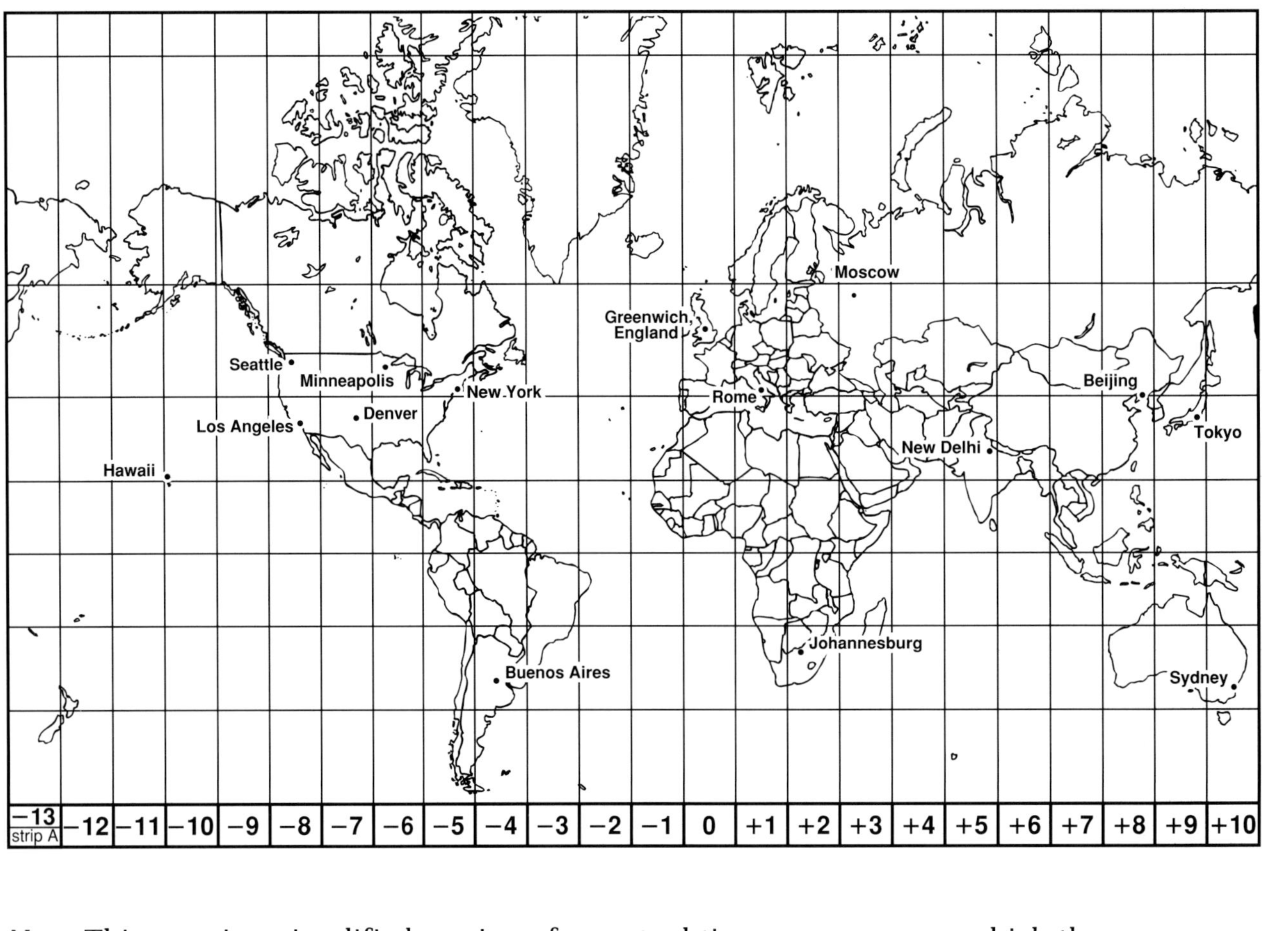

Note: This map is a simplified version of an actual time zone map, on which the zones often vary to accommodate islands, country borders, and certain geographic features. You may want to look at an atlas to see the actual time zones.

The continental United States has four time zones: eastern, central, mountain, and pacific.

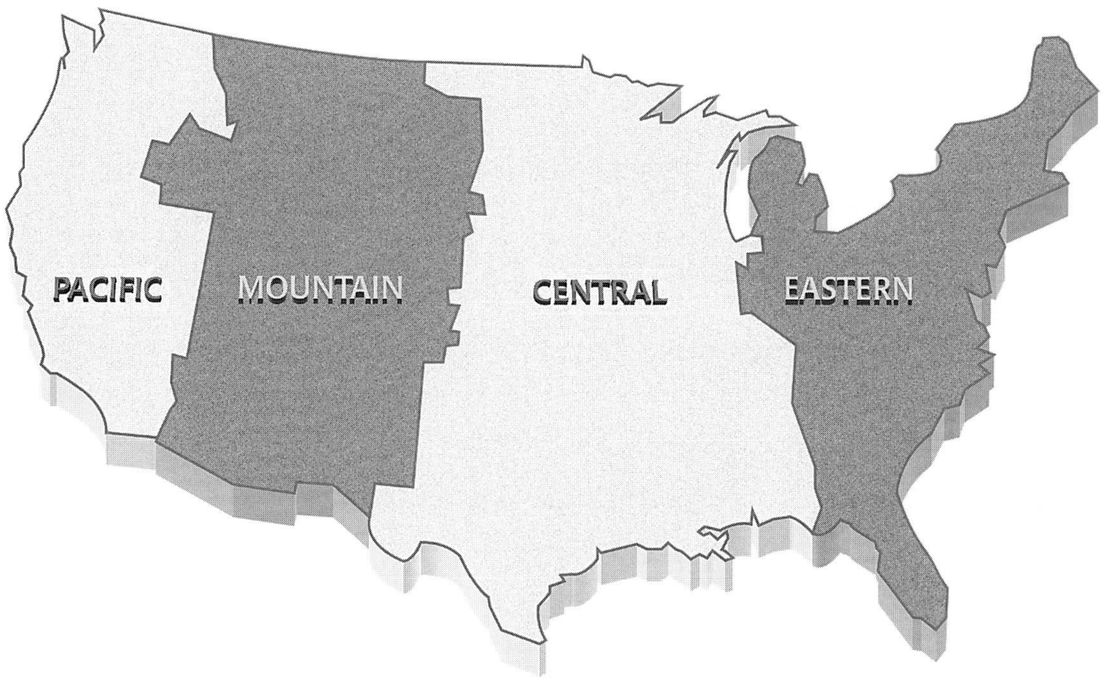

6. a. Compare the map above to your cylinder map from **Student Activity Sheet 2.** What numbers on the cylinder map correspond to the four time zones shown above?

b. If you travel from one time zone east to the next time zone on the cylinder map, what happens to the numbers?

c. When you travel east, should you change the time on your watch one hour backward or forward?

7. a. On the cylinder map, find the time zone in which you live. What is the number for your time zone? What does this number tell you?

b. In what time zone is Hawaii? What about Moscow?

c. Name a city in the time zone marked +2. What is the time difference between your time zone and the one in that city? Explain how you found your answer.

Tara, Victor, and José are classmates. The three students live in a city in time zone −5. They have relatives who live in countries all over the world. Tara's cousin Keisha lives in a place where it is 6 hours later than where Tara lives.

8. In what time zone does Keisha live? In what countries might Keisha live?

Victor's grandfather lives in time zone +5.

9. What is the time difference between where Keisha lives and where Victor's grandfather lives?

José's uncle lives in a place where it is two hours earlier than where José lives.

10. **a.** What is the time difference between where Keisha lives and where José's uncle lives?

b. How did you find this difference?

11. Keisha wants to phone Tara. Keisha's watch shows 4:00 P.M. Is now a good time to call? Why or why not?

Suppose that Tara lives in time zone +3.

12. **a.** If the time differences stay the same, in what time zone does Keisha live?

b. In what time zone does José's uncle live?

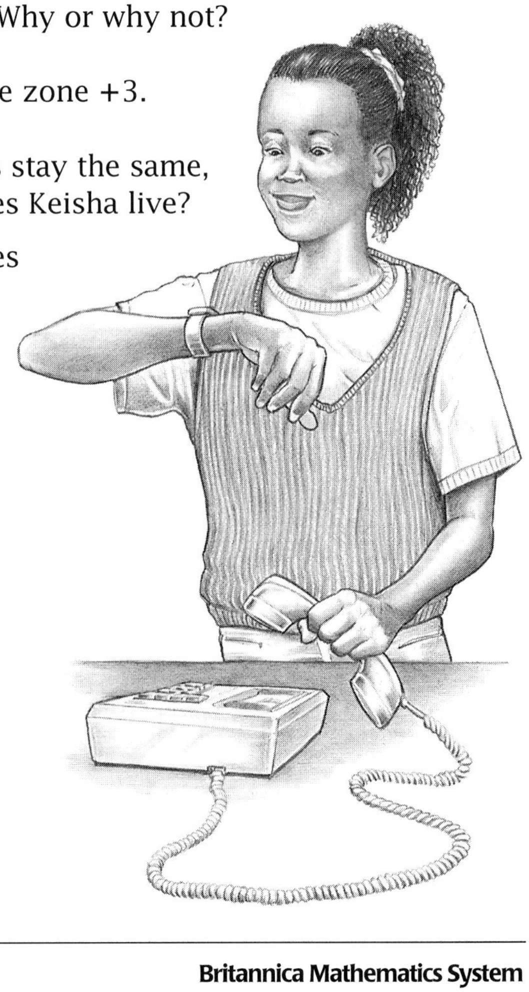

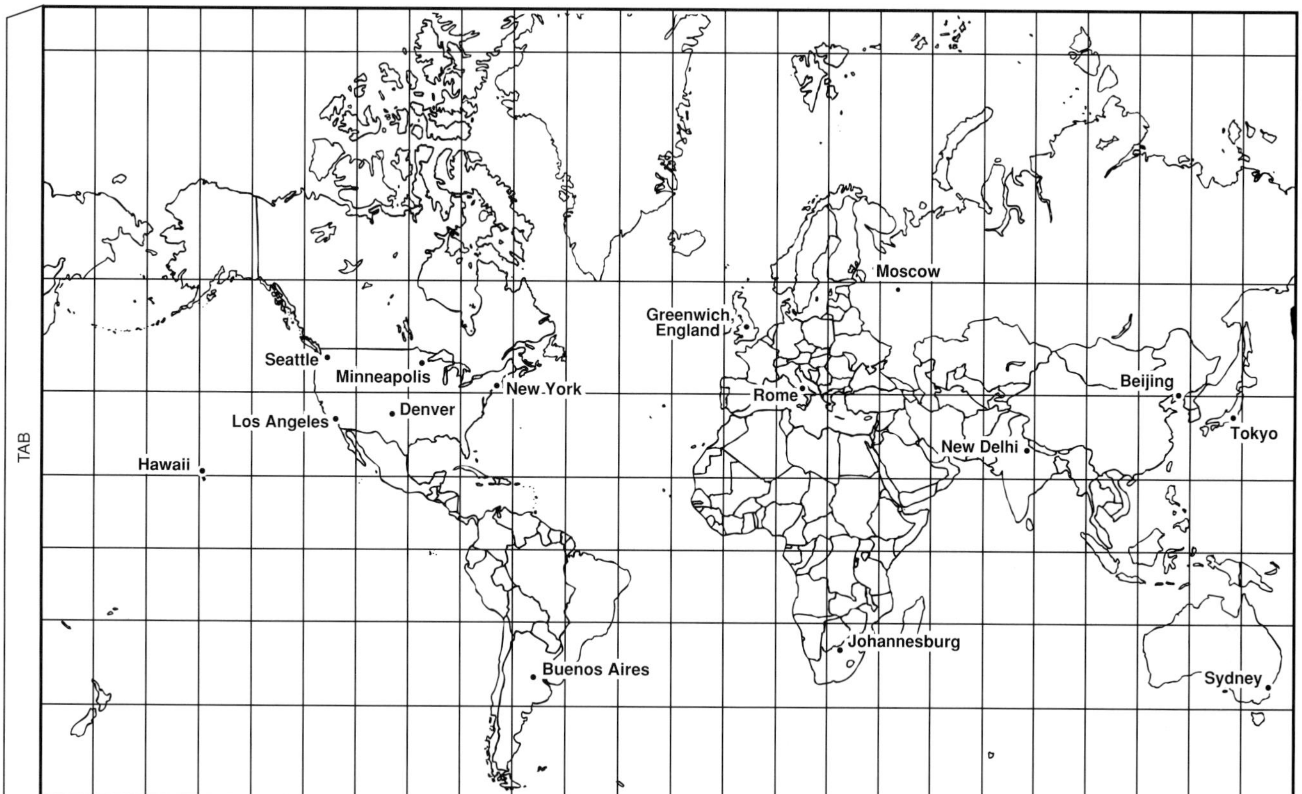

strip A	−13	−12	−11	−10	−9	−8	−7	−6	−5	−4	−3	−2	−1	0	+1	+2	+3	+4	+5	+6	+7	+8	+9	+10
strip B														−1	0	+1	+2							
strip C																								

For someone who lives in time zone 0, it is easy to calculate what time it is in other zones. Tony Catalano does not live in time zone 0. He lives in Rome. He makes many international phone calls, so he needs to know the times in other places in the world.

He decided to make a new strip that will be easier to use. Below strip A is the start of his new strip, strip B.

13. **a.** Why did he write a 0 under the +1 of strip A?

b. Fill in the rest of the numbers on strip B on a new copy of **Student Activity Sheet 2.**

14. In Rome, Tony's watch says that it's 10:00 P.M. Will he be able to reach a business associate in New York who works from 9:00 A.M. to 5:00 P.M.? Explain.

15. **a.** Choose a city on **Student Activity Sheet 2.** Write new numbers on strip C that will will make it easier to figure out the changes from that location.

b. Make up two time problems involving the place you choose for part **a** and solve them using your strip C. The time-zone problems might involve someone you know or an event from history.

Summary

In order to solve problems about time zones and time differences, you can calculate with positive and negative numbers.

Summary Questions

16. Read the story on page 12 again. Did the trip back from Seattle really take longer? Show how you found your answer.

17. Compare how you used positive and negative numbers in this section to show time changes with how you used positive and negative numbers in Section A to show moves in a design.

C. PLUS AND MINUS CHIPS

It's All in the Game

Mr. Jones's class is playing a game. The class is divided into four teams, and all four teams are given questions. Each team that answers its question correctly earns one plus point. Each team that answers incorrectly gets one minus point. The teams must answer all questions.

Renee and Sam are keeping score. Here is the score after four rounds.

1. How did each team get its points in the first four rounds?

On the left, you see the scores after teams A, B, and C have completed six rounds.

2. Explain how teams A, B, and C performed in rounds five and six.

Team D's score is being listed.

3. Why is it impossible for team D to have a total of +1 after round six?

4. Find all possible scores a team could have after six rounds.

On the left are the scores at the end of the game.

After they post the final score, the scorekeepers notice that they made a mistake: they gave a minus point to team B when team A really earned it.

5. Correct the scores for teams A and B. What is the correct final score for each team?

The students decide that they want to play the game again.

This time, Mr. Jones introduces a different way to keep score. He brings out some chips that have either a plus or minus symbol on them. He suggests that students keep score in the following way:

This seemed easy, so Gretta and Larry volunteer to be scorekeepers this time. They each get a stack of chips. After a number of rounds, each team wants to find out what the score is. Mr. Jones calls for a time-out to tally the current scores.

Below are the chips for teams A, B, and C.

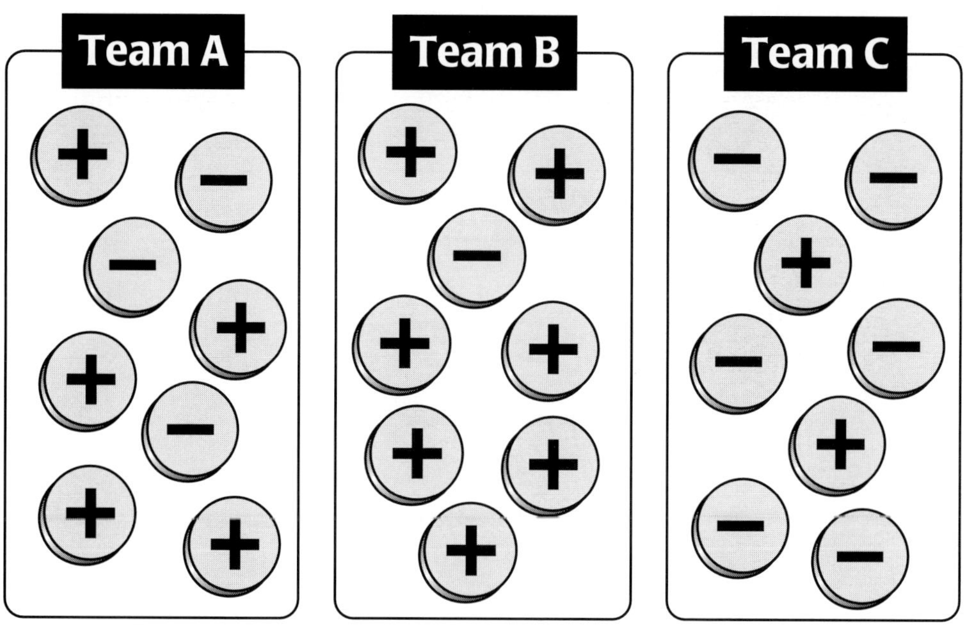

6. **a.** How many rounds have been played? How can you tell?

b. Find the score for each team.

Team D calculates its own score. Eun Young, the captain of the team, says

7. Draw a picture of the chips that team D has. Is there more than one possible combination of chips?

8. a. Describe or draw combinations of chips that add up to a score of +3 for different rounds in the game.

   b. Describe or draw combinations of chips that add up to a score of −3 for different rounds in the game.

Below are the chips for the four teams after 12 rounds.

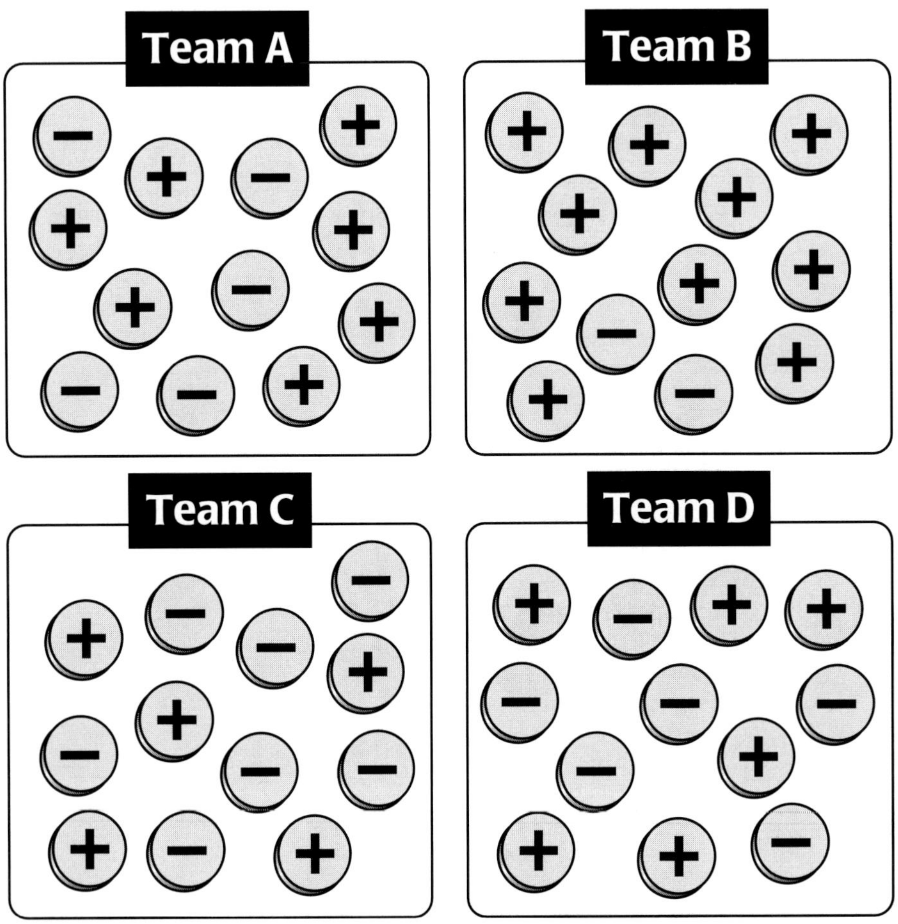

9. Find the score for each team.

10. The scorekeepers run out of "+" chips before the game ends. Find a solution to their problem.

11. The game continues for three more rounds. What are all the possible final scores?

During one game, a mistake is made. To correct the mistake, Gretta, the captain of team A, makes this suggestion to Larry.

12. Is the result the same? Explain your reasoning.

Below you see team B's chips after round 8.

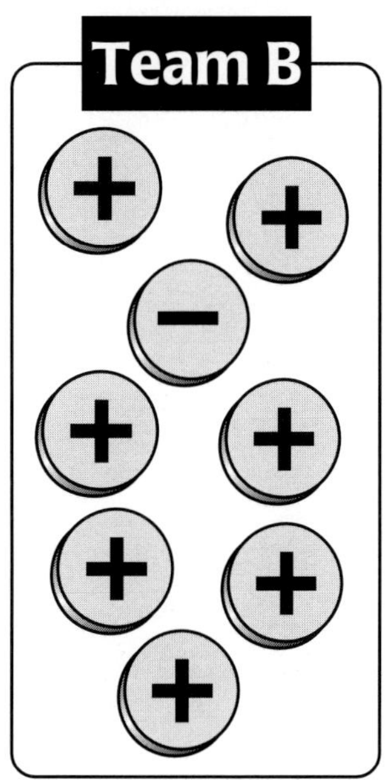

13. **a.** Take away some chips without changing the score.

b. Could you also add some chips without changing the score? Explain.

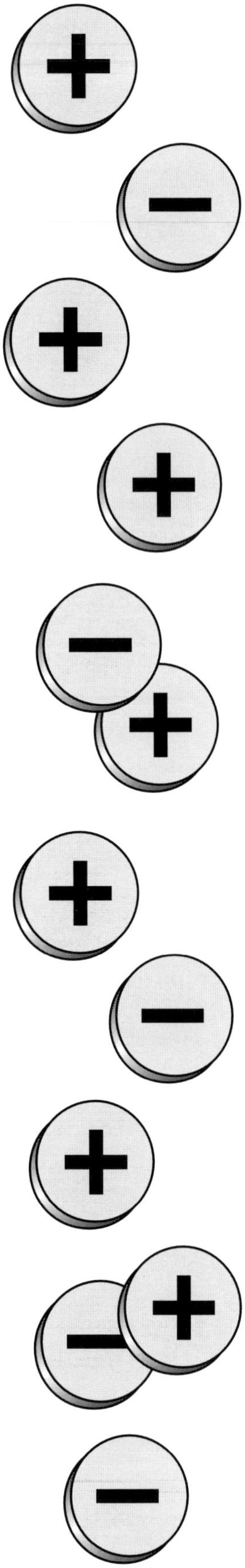

14. Suppose you are the scorekeeper for a game. What would you write down as the new score in each of the following situations? Copy and complete the chart below into your notebook. It may help to use chips or to draw pictures.

SCORE		POINTS	NEW SCORE
+7	add	+2	
+3	add	−4	
−5	add	+1	
−1	add	−2	
+6	take away	+3	
−1	take away	+5	
+2	take away	+8	
+5	take away	−3	
+4	take away	−6	
−6	take away	−2	
−3	take away	−8	

In problem **14** you found the following:

$$+5 \text{ take away } -3 = +8$$

15. a. Instead of "take away −3," what can you do to get the same new score? Rewrite the problem by replacing "take away −3."

b. Copy the problems below and fill in the underlined parts.

+5 take away −3 = +8 gives the same new score as +5 ____ = +8

−3 take away −8 = ____ gives the same new scores as −3 ____ = ____

Instead of the word "add," you can use the addition sign (+). Instead of the words "take away," you can use the subtraction sign (−).

16. Rewrite the calculations from problem **14** with + and − signs. Be careful not to mix up the signs!

17. Copy the following equations. Write numbers (both positive and negative) in the blanks to make the equations true.

a. ____ + ____ = +2 **b.** ____ + ____ = +1

c. ____ + ____ = 0 **d.** ____ + ____ = −1

e. ____ − ____ = +2 **f.** ____ − ____ = −3

g. −1 − ____ = 0 **h.** −3 − ____ = +2

Summary

In this section, you added and subtracted positive and negative numbers.

Summary Questions

18. **a.** Explain what happens when you add a negative number to another number. It might help to use examples like $+4 \ + -1 =$ ____ and $-3 \ + -1 =$ ____, or to think of scores or chips.

b. Use scores or chips to explain what happens when you subtract a negative number from another number.

19. Which of these statements are *always* true? Explain your answers.

a. A positive number added to a positive number results in a positive number.

b. A positive number added to a negative number results in a negative number.

c. A negative number subtracted from a positive number results in a positive number.

d. A negative number added to a positive number results in a negative number.

e. A negative number added to a negative number results in a negative number.

D. ALTITUDES AND TEMPERATURES

Altitudes

Diego and his mother are hiking in the mountains in Colorado. They hike to a summit that is about 600 meters higher than their town.

Here is Diego's story of the trip to the summit.

The summit was very high, so we had to go uphill a lot. There were also times when we had to go downhill. When we were at the top, we were much higher than where we started.

1. How is it possible that Diego and his mother had to go downhill at times when hiking to the summit? Illustrate your answer with a drawing.

When Diego returned to school, he told his class about his hiking trip in Colorado. On one trail, the ending point is 150 meters lower than the starting point.

2. Make a drawing that shows what the trail might look like.

The picture below is from a Colorado hiking guide. It is a drawing of a trip from Clear Creek Road to North Cottonwood Creek Road.

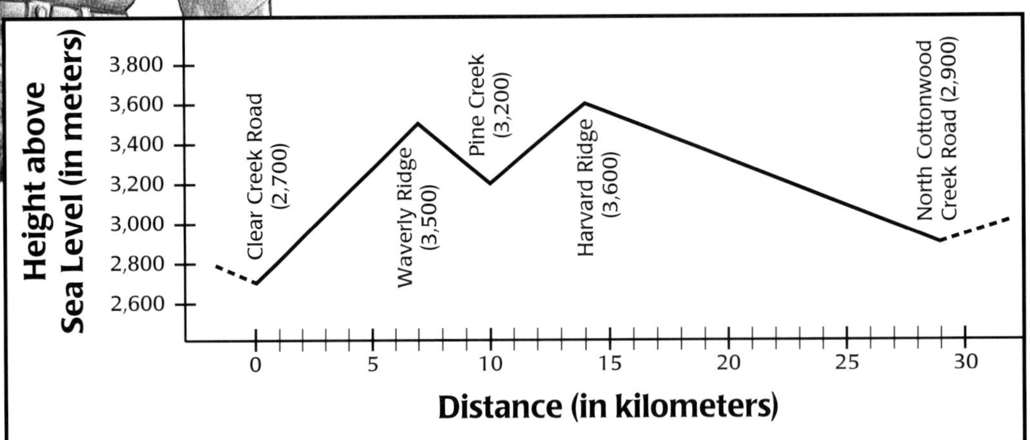

3. a. Describe the above trip in words. Do you think this is a difficult trip?

b. Does the trail end higher or lower than its starting point? by how many meters?

4. a. How many meters higher is Waverly Ridge than Clear Creek Road?

b. When hiking the trail, how many meters do you climb and how many meters do you descend?

Carlita loves to hike in the mountains. When she plans a trip, she uses drawings from her hiking guide. Below you see a drawing from her book.

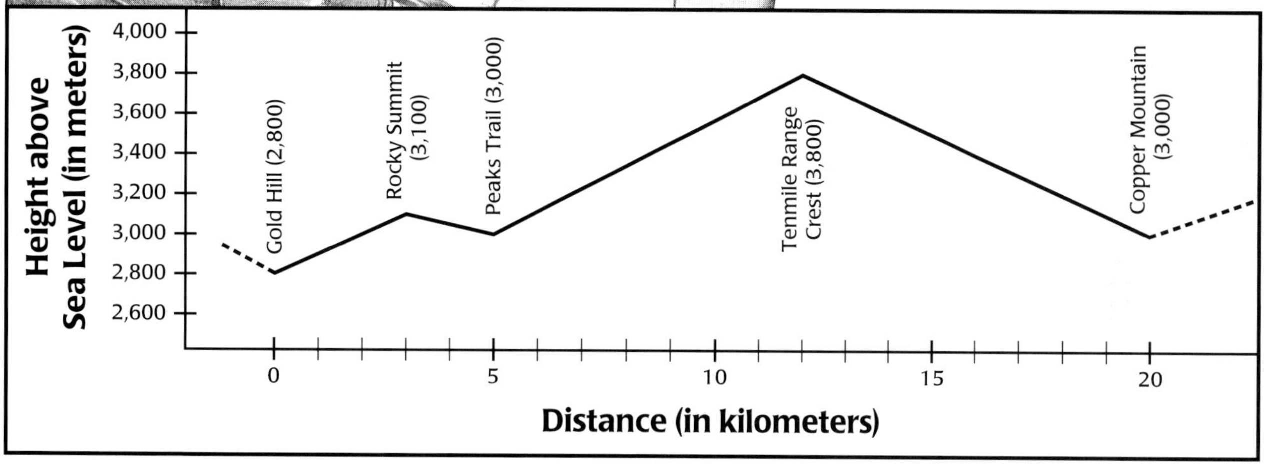

Carlita writes the following numbers on the drawing.

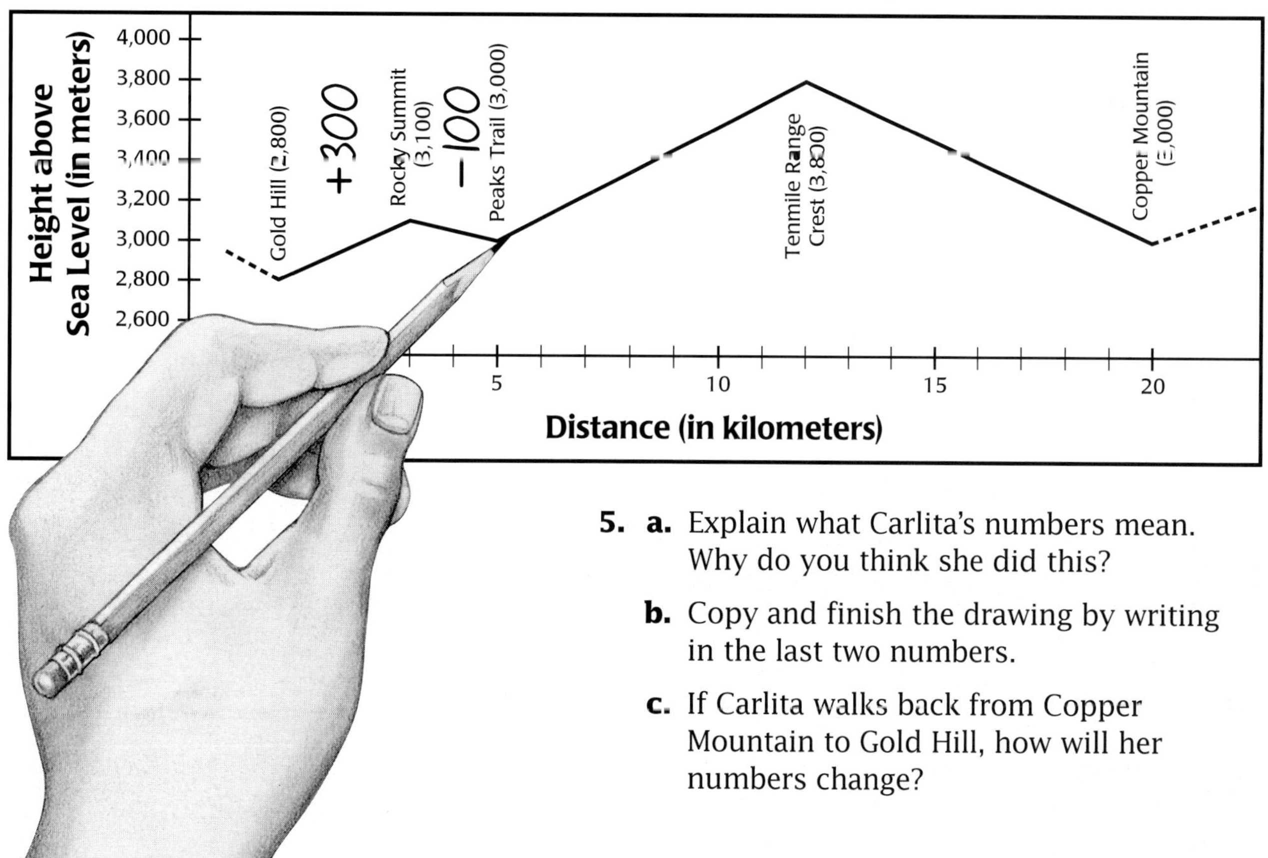

5. **a.** Explain what Carlita's numbers mean. Why do you think she did this?

b. Copy and finish the drawing by writing in the last two numbers.

c. If Carlita walks back from Copper Mountain to Gold Hill, how will her numbers change?

Bat Cave Trail

Uphill (in meters)	Downhill (in meters)
+125	−300
+200	−125

Another way to show how many meters a trail goes up and down is with a table as shown on the left.

6. **a.** Is the endpoint of Bat Cave Trail higher or lower than its starting point?

b. Draw Bat Cave Trail so that it fits the numbers in the table. Compare your drawing to a classmate's. How can you mark the trail so that it is easy to find the height differences?

Mirror Lake Trail

Uphill/ Downhill (in meters)
+230
−130
+37
−340
+110
−37
+140
−40

On the left is the table for Mirror Lake Trail.

7. Estimate whether you would end higher or lower than where you started if you hiked this trail.

Jassir uses the following method to find out exactly how many meters higher or lower the endpoint is:

8. Find the height difference between the endpoint and the starting point for Mirror Lake Trail using Jassir's method.

Trail A: +215, −115, +245, −300, +55, −75, +115

Trail B: +170, −335, +230, −500, +235, −65

9. For each of the trails on the left, find out how much higher or lower the endpoint is than the starting point.

Temperatures

Mark lives in Colorado. One week during the winter, he decides to record the high and low temperatures for one week.

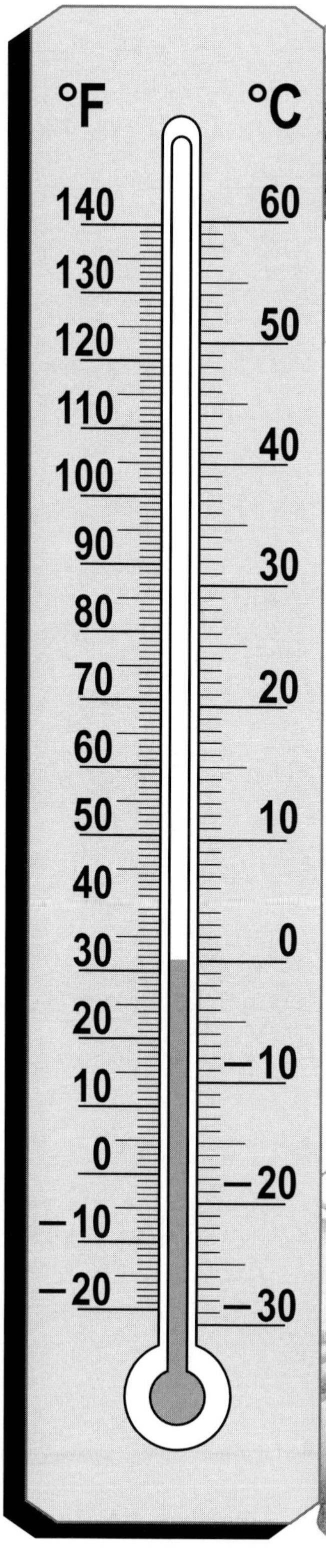

Here are the temperatures Mark recorded.

	Sun	Mon	Tue	Wed	Thu	Fri	Sat
High Temp (°C)	9	3	4	−2	−1	3	−1
Low Temp (°C)	−1	0	−4	−3	−7	−1	−2

10. a. Which day was the coldest? Which day was the warmest? Explain your answers.

b. Write the low temperatures in order from coldest to warmest. Was the temperature below freezing every day?

11. Calculate the mean high temperature for the week. What was the mean low temperature? Show your calculations.

	Sun	Mon	Tue	Wed	Thu	Fri	Sat
High Temp (°C)	9	3	4	−2	−1	3	−1
Low Temp (°C)	−1	0	−4	−3	−7	−1	−2

On Sunday, the high temperature was 9°C, and on Monday it was 3°C. To show that it was warmer on Sunday than on Monday, you can write $9° > 3°$ and say "Nine degrees is more than three degrees."

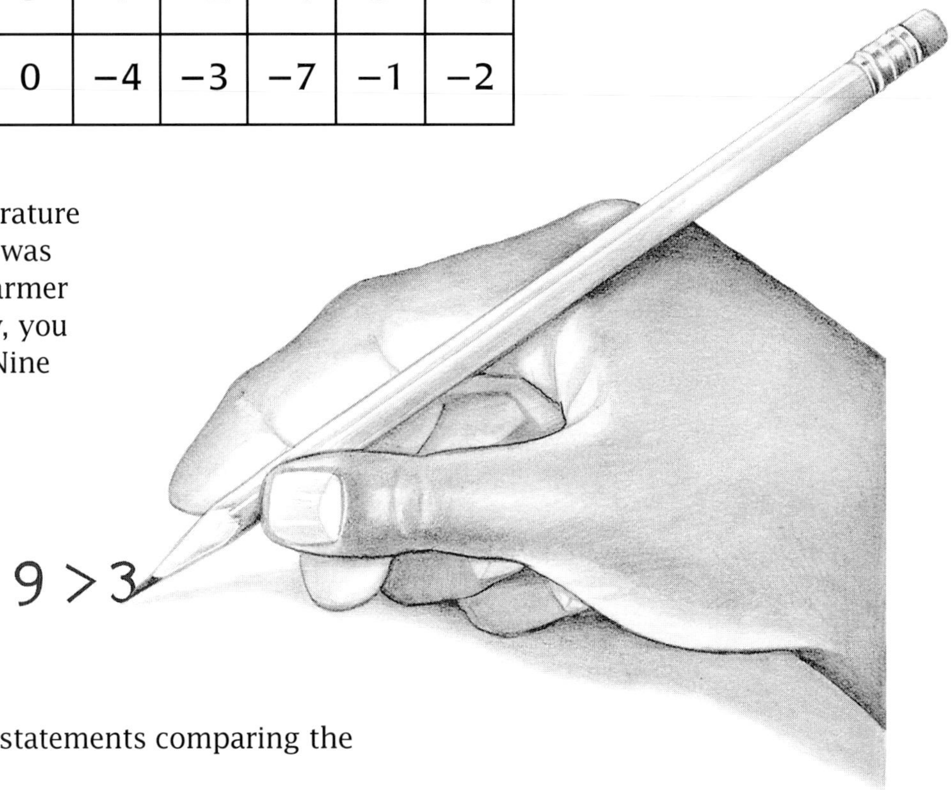

12. Write some other true statements comparing the high temperatures.

The high temperature on Thursday was lower than the high temperature on Friday. You can write $-1° < 3°$ and say, "Minus one degree is less than three degrees."

13. Is it also true that Thursday's low temperature was less than Friday's low temperature?

14. Which of the statements below are true? Explain in words what each statement says using Mark's temperature table.

a. $4 > -2$ **b.** $-2 > -1$ **c.** $-1 < 3$ **d.** $3 < -1$

15. Compare each high temperature in the table above to the mean high temperature that you found in problem **11** and write a true statement using $<$ or $>$.

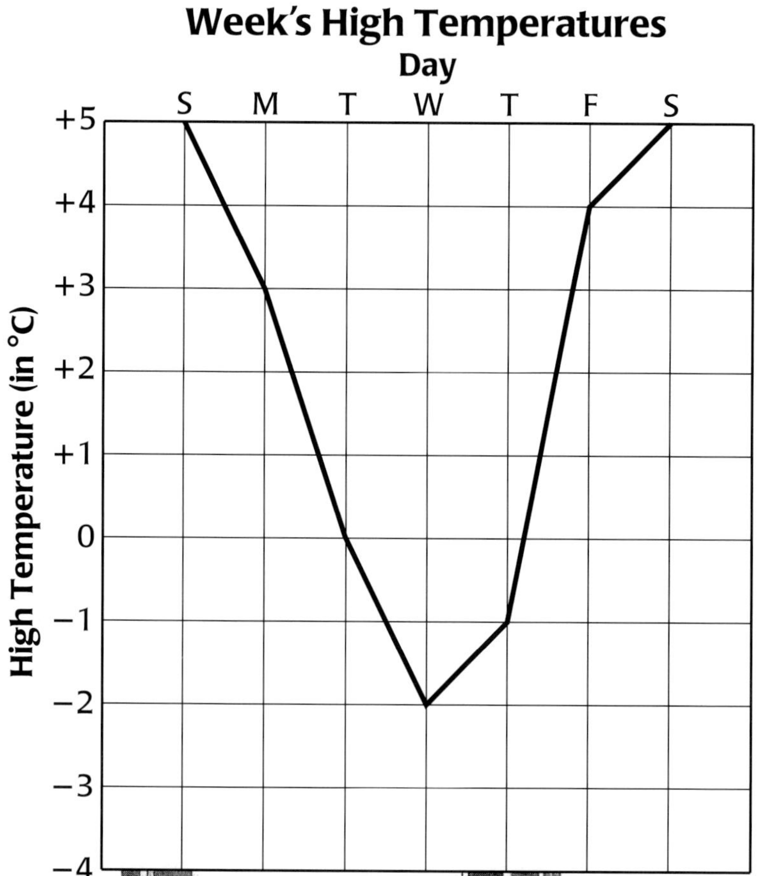

The next week, Mark again keeps records of high and low temperatures. Instead of writing them in a table, he makes the graph on the left. This graph shows the high temperature for each day.

16. Compare the high temperatures from this graph with those from the table for the previous week. Which week was warmer? Give reasons to support your answer.

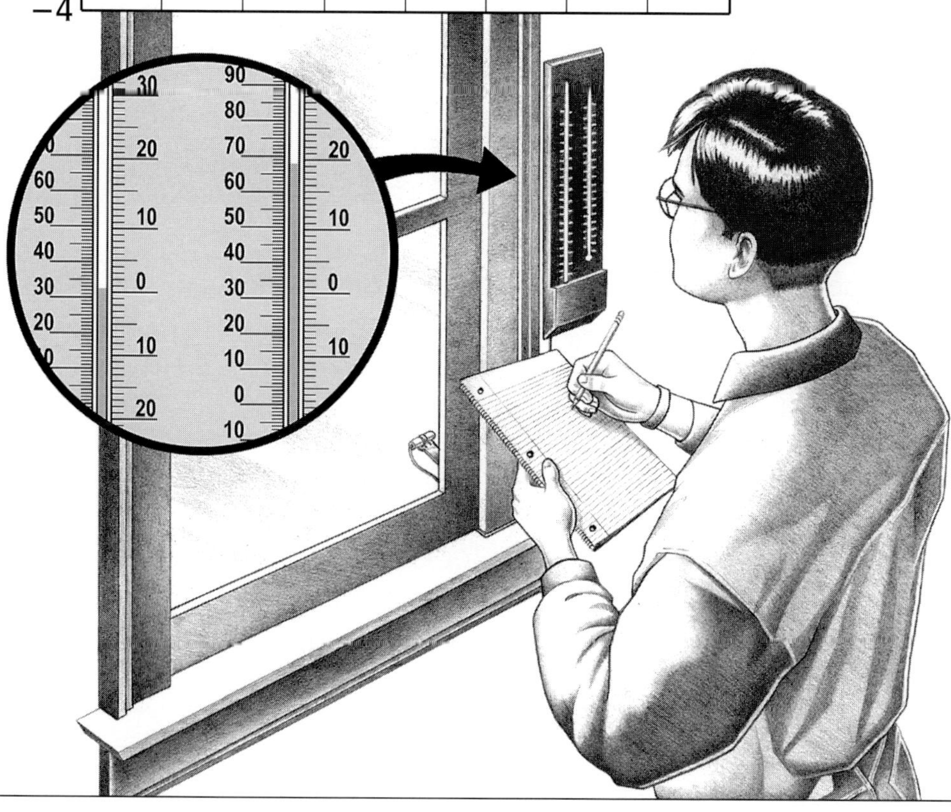

Mark calculates the mean low temperature of the second week to be −1°C.

17. Come up with one possibility for each of the seven daily low temperatures that week.

Here is a list of high temperatures for the whole month of January.

Sun	Mon	Tue	Wed	Thu	Fri	Sat
	−1	+4	0	−1	−1	+1
−2	+3	0	0	−1	+1	+1
−3	−3	−5	−1	0	0	+1
+1	+2	+2	0	−2	−2	−1
−1	+2	+2	−2			

Mark and his sister Tracy both want to calculate the mean high temperature for January. Tracy starts by adding all of the temperatures.

Mark starts by tallying how often each temperature occurred.

−5	−4	−3	−2	−1	0	+1	+2	+3	+4
				///	/	/			/

18. a. Explain how both Tracy and Mark should continue in order to calculate the mean.

b. What method do you like best? Why?

19. a. Copy and finish Mark's tally table.

b. Because there are four days with a temperature of −2°C, Mark calculates

$$4 \times -2 = -2 + -2 + -2 + -2 = \_\_\_$$

What is the answer? Make calculations like this for each entry in the table.

c. Calculate the mean high temperature.

HIGHER and HIGHER

2,300 meters

Mark and Tracy live near a ski area. There are three ski lifts on the mountain. The lowest one is at 1,300 meters above sea level. From this lift station there is a ski lift that can take you to the highest lift station at 2,300 meters. There is also a middle lift station halfway between the other two stations.

A sign at the lower station shows the temperatures at each ski-lift station at different times of the day.

1,800 meters

Altitude (in meters)	Ski-Lift Station	Temperature (°C) at		
		8:00 A.M.	Noon	6:00 P.M.
2,300	Upper	-8		
1,800	Middle			
1,300	Lower	-2		

20. a. What do you think the temperature at the middle lift station was at 8:00 A.M.?

b. What do you notice about the temperature as you move higher up the mountain?

1,300 meters

Altitude (in meters)	Ski-Lift Station	Temperature (°C) at 8:00 A.M.	Noon	6:00 P.M.
2,300	Upper	-8		
1,800	Middle			
1,300	Lower	-2		

At noon, the temperatures at all three ski-lift stations have risen by 5 degrees.

21. **a.** On a separate sheet of paper, write down the noon temperatures for all of the ski-lift stations.

b. Assume that the differences in temperatures between the ski-lift stations are always the same. What might the temperatures be at 6:00 P.M.?

In the mountains, for each change in height there is a corresponding change in temperature.

22. In Mark and Tracy's ski area, what happens to the temperature each time you move up 500 meters?

The next day at noon, the temperature at the middle ski-lift station is −1°C.

23. At noon, what temperatures will the sign show for the following places?

Altitude (in meters)	Place	Temperature (°C)
3,300	Summit	
2,300	Upper Lift Station	
1,800	Middle Lift Station	-1
1,300	Lower Lift Station	
1,050	Valley Village	

Here are the high temperatures for a week at the middle ski-lift station.

Day	Sun	Mon	Tue	Wed	Thu	Fri	Sat
High Temp (°C)	−5	−3	−3	0	−4	−3	−5

24. What was the mean high temperature that week at the upper lift station? at the lower lift station?

The relationship you discovered between temperature and altitude tends to be true all over the world.

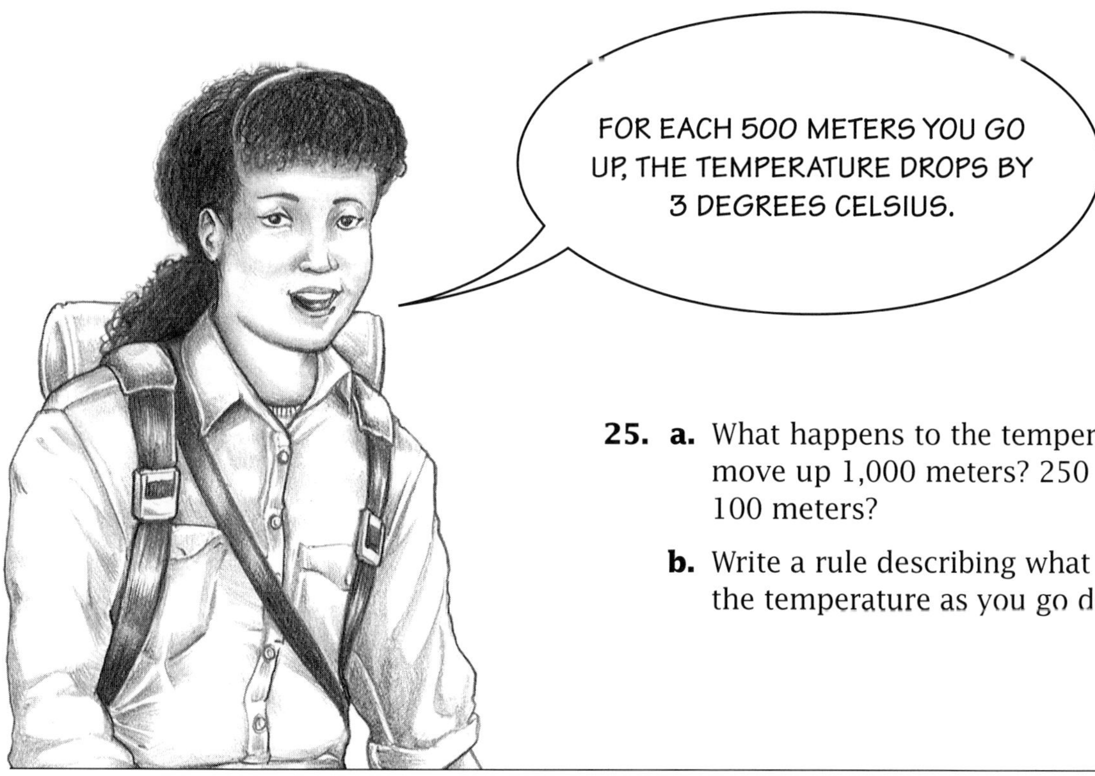

25. a. What happens to the temperature if you move up 1,000 meters? 250 meters? 100 meters?

b. Write a rule describing what happens to the temperature as you go down.

You may find heavy snow at higher altitudes. Sometimes snow can remain at high altitudes all summer long.

Suppose that there is a mountain whose summit is 4,418 meters above sea level. This mountain descends into the sea. At sea level, the temperature is 23°C.

26. Is the snow melting at the summit? You may use a table like the one below to help you.

Altitude (in meters)	0 (sea level)	500	1,000	
Temperature	23°C			

Marcos is working on the following problem:

> The temperature at the foot of a mountain is 12°C. What is the temperature 300 meters up?

Marcos writes down 12 + 3 × −0.6 =

27. Explain how this calculation fits the problem and then find the answer.

Look again at the trail Carlita plans to take. She knows the rule about temperatures and height. When Carlita starts at Gold Hill, it is about 22° C.

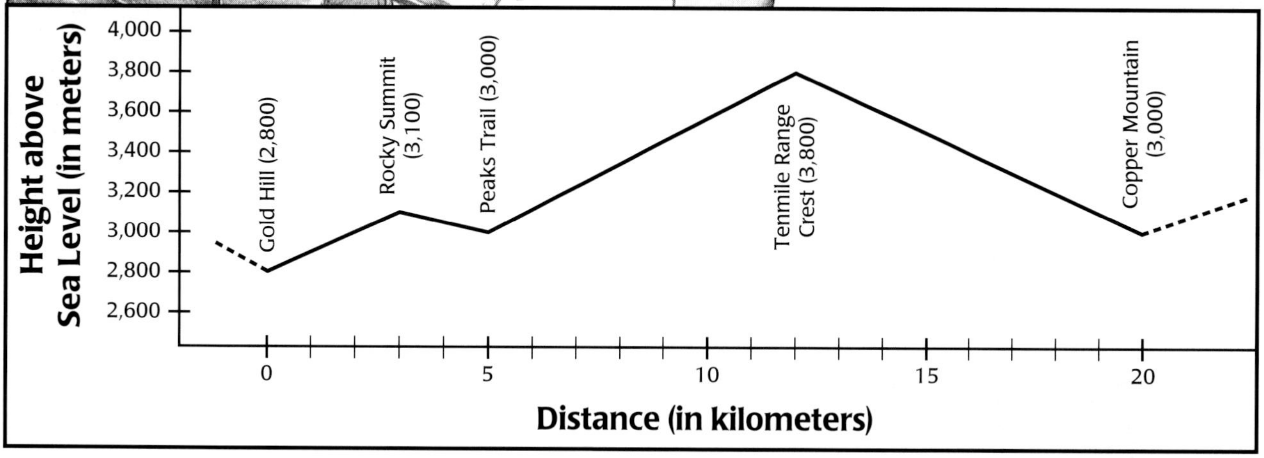

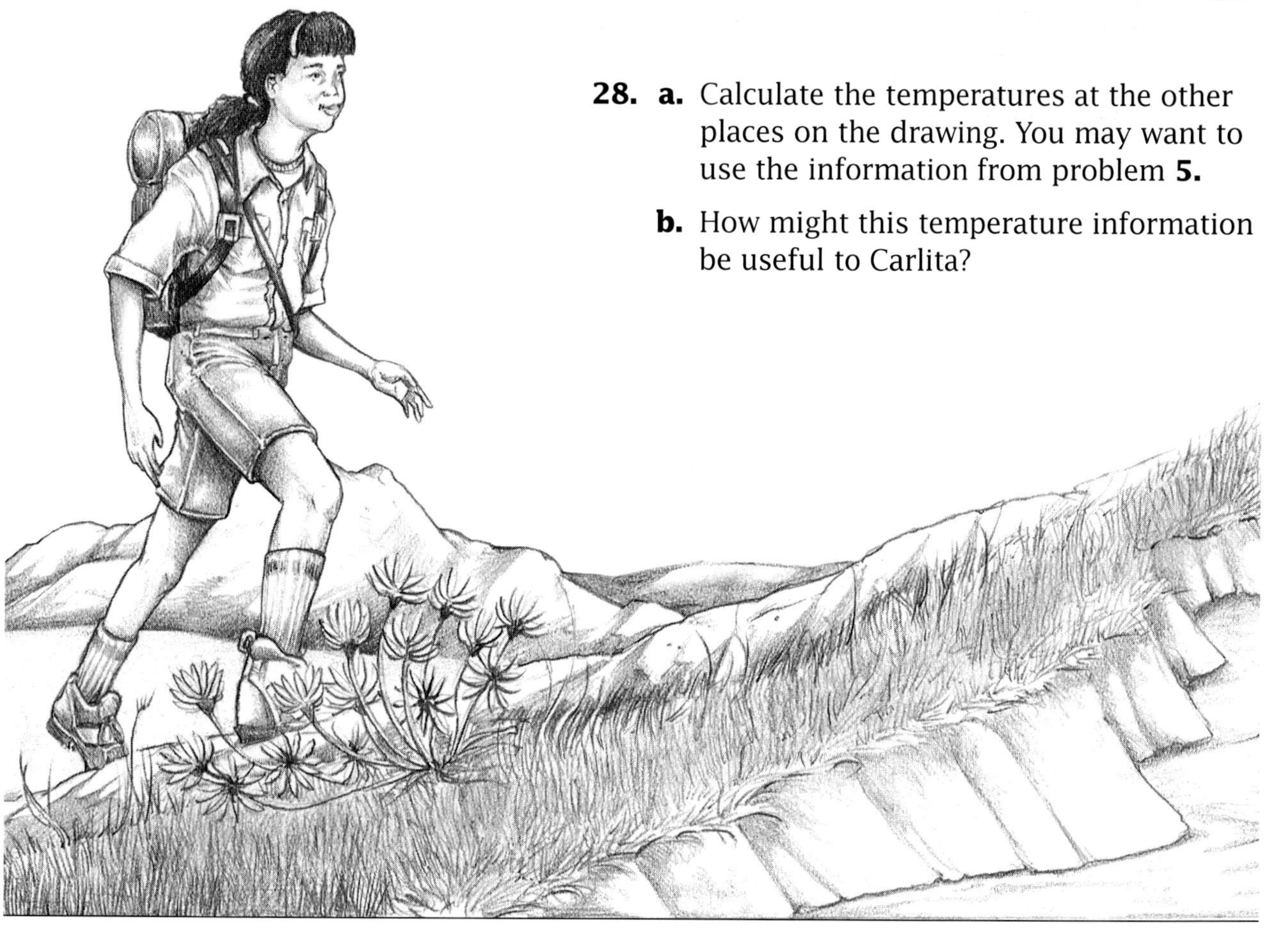

28. **a.** Calculate the temperatures at the other places on the drawing. You may want to use the information from problem **5.**

b. How might this temperature information be useful to Carlita?

Summary

Did you notice that positive numbers are sometimes written with a + in front of the number and sometimes without? Either way, they mean the same thing. However, you cannot leave out a negative sign when writing a negative number.

There are different ways to add a group of positive and negative numbers:

- You can add them in their given order.
- You can add all of the positive numbers, add all of the negative numbers, and then add the two results.
- You can first add the numbers that are easily combined, such as 65 and 35, or 24 and −24.

You can also multiply and divide positive and negative numbers.

The symbols < and > can be used to compare two numbers: $a < b$ means that a is less than b, and $b > a$ means that b is greater than a.

Summary Questions

29. For 10 days, high temperatures were recorded. On four days it was −3°C; on three days it was −2°C, on one day it was −1°C, and on two days it was +2°C.

a. Explain why you can begin to find the mean temperature as follows:

$$(4 \times -3) + (3 \times -2) + (1 \times -1) + (2 \times 2) =$$

b. Finish the calculation for the mean temperature.

30. Here are some practice equations for finding means. For each one, complete the calculation.

a. $-100 \div 25 =$ **b.** $-31 \div 31 =$

c. $40 \div 12 =$ **d.** $-5 \div 10 =$

31. Make up a story problem that involves the multiplication of a positive and a negative number.

32. Write a story problem that requires one of the calculations in problem **30.**

E. ALL OPERATIONS

Positive and Negative Numbers

In Section D, Tracy and Mark tallied temperatures and then calculated to find the mean.

1. Make up a story about temperatures for the following calculation:

$$2 \times -1 + 3 \times -2 + 2 \times 2$$

2. Make up a story about temperatures for the calculation $-42 \div 12$. Then find the result.

In the two examples above, you divided or multiplied with positive and negative numbers.

3. Determine whether each of the statements below is true for all numbers, some numbers, or no numbers.

a. positive × positive = positive

b. positive ÷ positive = negative

c. positive × positive = negative

d. negative ÷ positive = negative

e. positive × negative = negative

4. Write down one other true statement and one other false statement about calculating with positive and negative numbers.

In Section C, you used chips to find scores with positive and negative points. A plus chip represented +1 point and a minus chip represented −1 point.

5. How would you use the plus and minus chips to show what 4×-2 means in terms of points?

Multiple **Choice**

Mr. Jones is going to give his class a multiple-choice worksheet on some vocabulary words. He doesn't want students to simply guess which of the given definitions is correct, so he decides to use the following system for earning points.

He will give 4 points for every correct answer, 2 minus points for every incorrect answer, and 1 point for a blank. There are 20 questions on the worksheet.

6. a. What is the lowest score a student can get?

b. What is the highest score a student can get?

7. a. Suppose Sally gets five questions wrong. How many minus points will she have? Show how you calculated your answer.

b. If Sally answers all of the other questions correctly, how many total points will she have? Show your calculations.

8. Suppose Theo gets only three questions wrong, but he leaves two blank. What is his score? Show your calculations.

Mr. Jones discovers that there is a mistake in the scoring on John's paper. Two answers were marked correct that are really incorrect.

Mr. Jones writes $(-2 \times 4) + (2 \times -2)$ on top of John's paper.

9. a. Explain why Mr. Jones does this.

b. How does John's score change? Explain your answer.

Paul is concerned because the score on his worksheet is 50. When Mr. Jones looks at the worksheet, he finds some mistakes in scoring. Mr. Jones recalculates Paul's score by counting three problems correct that he had marked incorrect.

10. a. Write a number sentence for the calculation. What is the answer?

b. What is Paul's final score?

Chips Again

For problems **11–13,** you may want to use a pile of chips.

Start with 12 plus chips and 12 minus chips.

11. a. What is your total score?

b. Take away three groups of four positive chips. Write a number sentence describing this.

c. Describe how you could show -3×-4 with plus and minus chips. What is -3×-4?

12. Compare the results of 3×-4 and -3×4. You may use chips if you like. Do you get different answers?

13. Complete the following multiplications. You can use chips to help you.

a. $+4 \times -6$ **b.** -3×-5

c. -1×-3 **d.** $-5 \times +7$

e. $+4 \times +8$ **f.** $-2 \times +14$

Summary

Now you should be able to add, subtract, multiply, and divide with positive and negative numbers. You can use chips to help you find the answers. When you multiply and divide, there are different rules depending on whether the numbers are positive or negative.

When you multiply a negative number and a positive number, changing the order of the positive and negative numbers does not change the answer. For example, -2×3 gives the same result as 3×-2.

Summary Questions

14. Copy and fill in the blanks for the following rules:

If you multiply two positive numbers, the result will be ________.

If you multiply a positive number and a negative number, the result will be ________.

If you multiply two negative numbers, the result will be ________.

15. Copy the tables below. Some results have already been calculated. Figure out how the tables work and fill in the other answers.

−	−2	−1	0	+1	+2
−2					
−1	+1				
0				−1	
+1				0	
+2					

×	−2	−1	0	+1	+2
−2					
−1					
0					
+1					+2
+2					

F. OPERATIONS AND COORDINATES

The Coordinate System

To communicate about the locations of points on a grid, it is useful if everybody uses the same language and notation. A grid with a fixed starting point can be helpful. This point is called the *origin*, or point *O*.

When you plot on a grid with an origin, every point is located in reference to point *O*.

On the grid below, point *F* lies two to the right and one down from the origin. The location of this point is written as $F(+2, -1)$. These numbers are called the *coordinates* of point *F*.

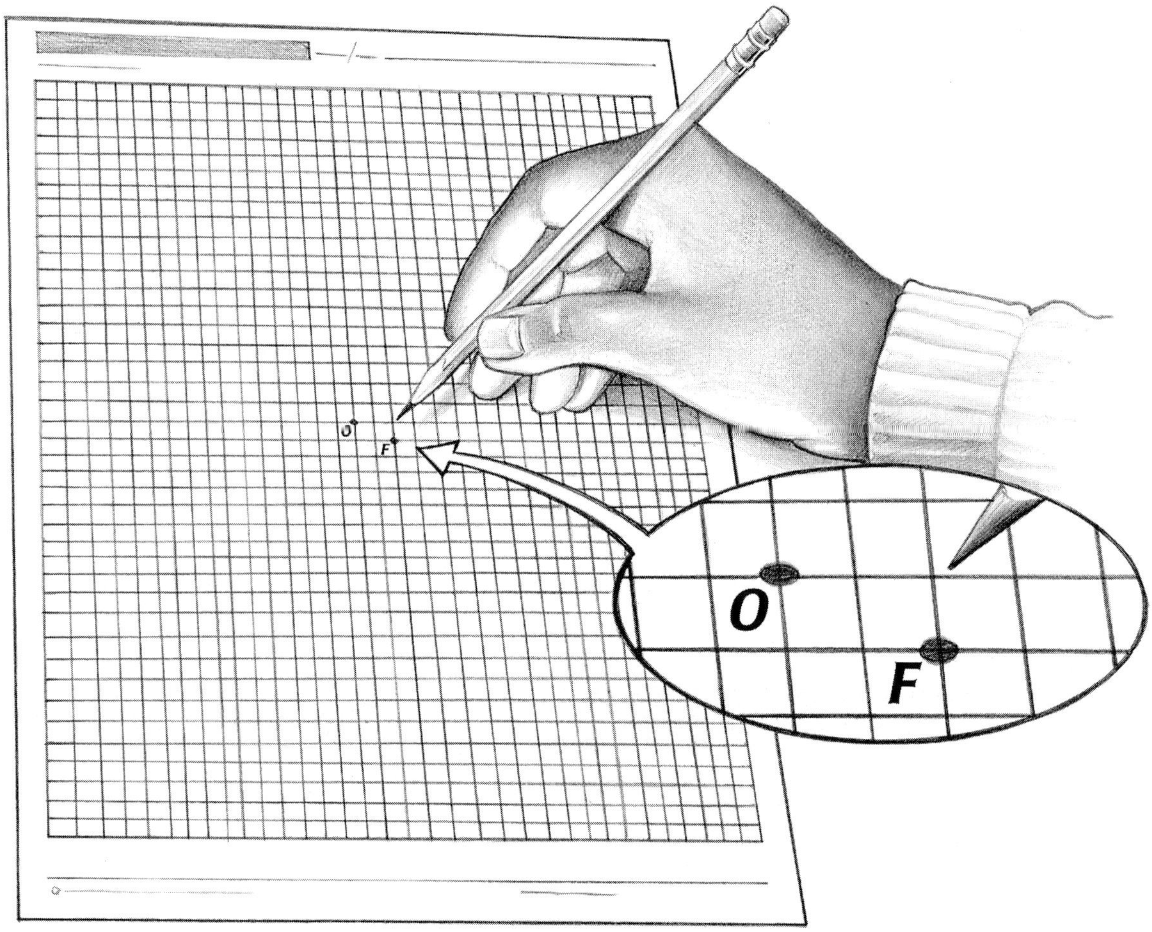

1. The first coordinate of *F* is positive. Why is the second coordinate negative?

2. a. Put point $P(+2, +1)$ on the grid on **Student Activity Sheet 3.** (Remember to start at O.) How would you describe the location of P?

b. Add $Q(-2, +1)$, $R(-6, -4)$, and $S(0, +3)$ to the grid.

c. Is $(-2, +1)$ the same as $(+1, -2)$? Explain your reasoning.

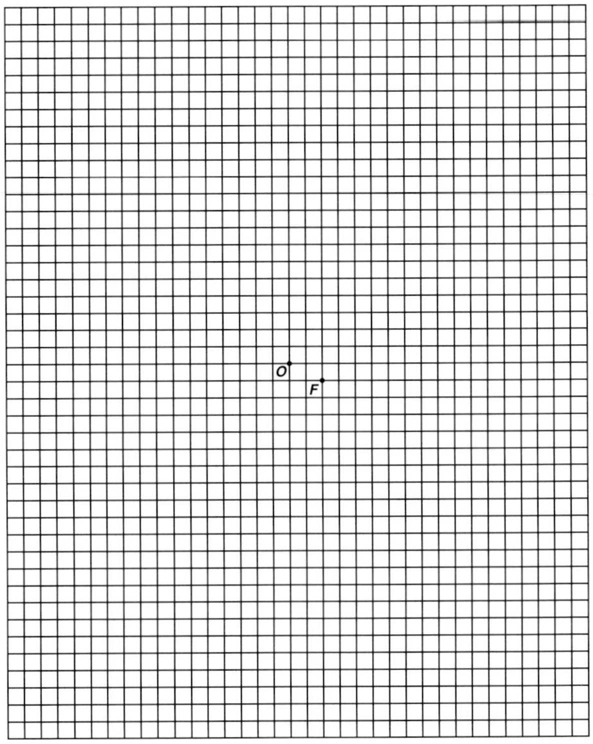

You probably noticed that you have to count each time you want to find the coordinates of points. Some grids have numbers to help you. A grid like this is called a *coordinate system*, and it has one number line going across and another going up and down. The line going across is called the *horizontal axis*, and the line going up and down is called the *vertical axis*.

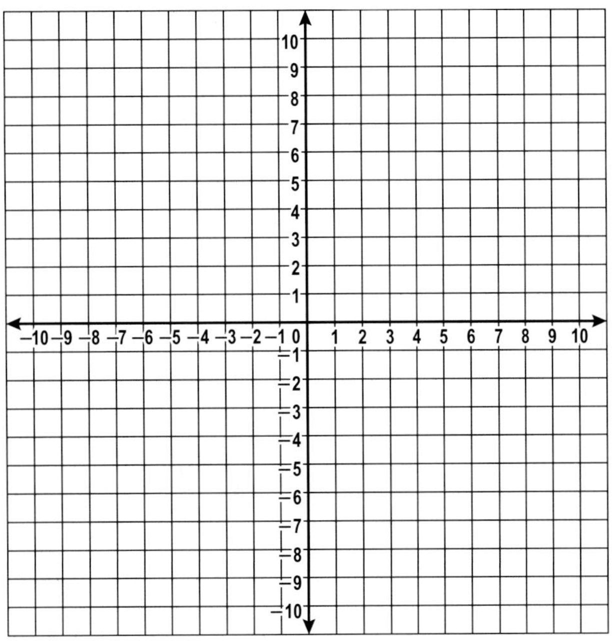

Use **Student Activity Sheet 4** for problems **3–9.**

3. a. Draw vertical and horizontal axes. Put a number scale on each axis and a 0 for the origin. (We changed the O to 0; why do you think we did that?)

b. Plot and label the following points on your coordinate system: $A(1, 1)$, $B(3, 1)$, $C(3, 3)$, $D(2, 4)$, and $E(1, 3)$.

c. Connect the points to make a polygon.

Now you are going to do some operations on the coordinates of *A, B, C, D,* and *E* from problem **3.**

For problems **4–8,** first predict what result each operation will have on the polygon. Then check your prediction by making a drawing. For each problem, start with the coordinates given in problem **3.**

4. Add −3 to the first coordinate of each point. What happens?

5. Add 1 to the first coordinate and −5 to the second coordinate of each point. What happens now?

6. What happens if you subtract 1 from all of the coordinates?

7. What should you do to the coordinates if you want to move the figure up three squares?

8. What happens to the figure if you multiply both coordinates of each point by 3?

Consider the following conversation between Cassandra and Paul.

9. **a.** Find the area of the figure you made for problem **3.**

b. Find the area of the new figure you made in problem **8,** and compare it to the area of the figure in problem **3.** How much larger has the figure become?

c. Does the result surprise you? Why or why not?

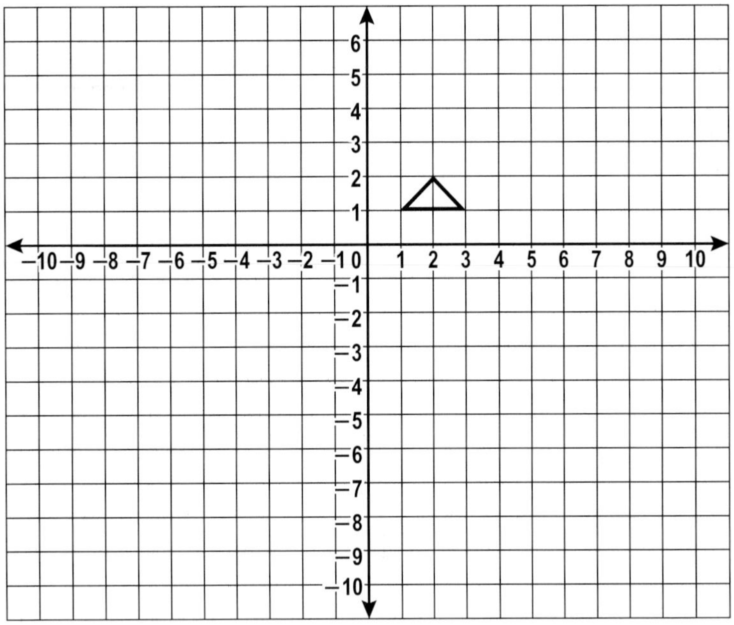

10. **a.** On a new copy of **Student Activity Sheet 4,** make a new coordinate system. Draw the triangle on the left.

b. What are the coordinates of each vertex?

c. Multiply all of the coordinates by −3. What are the new coordinates of the vertices?

d. Draw the new figure in the same coordinate system.

Carrie wonders what would happen to the figure she made in **10d** if she multiplied the coordinates by −3 again. This is what some of her classmates think.

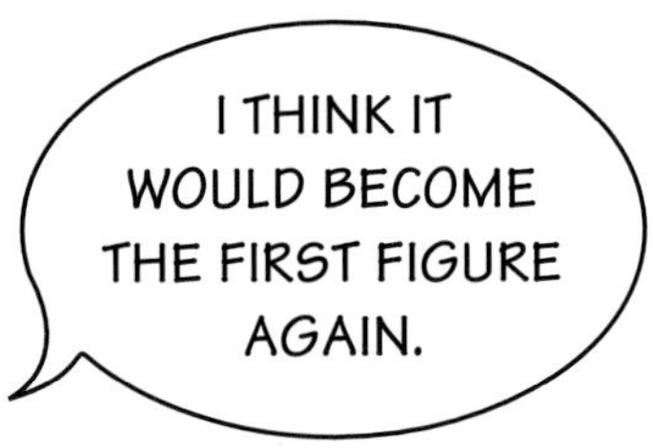

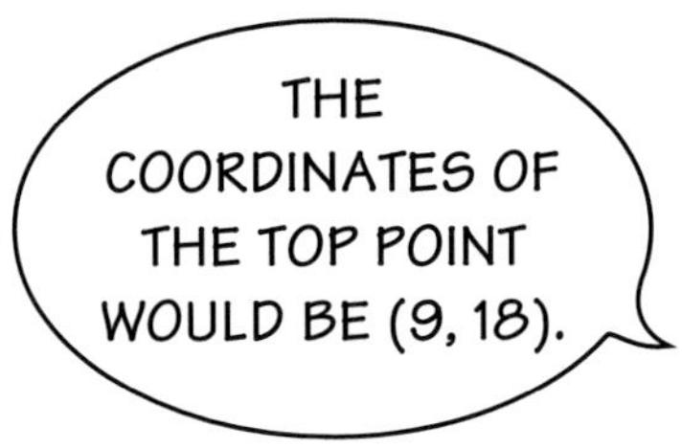

11. **a.** Who is correct? Why?

b. Write down everything you know about what will happen to the figure.

Gil, Lashonda, and Greg are discussing how they might shrink a triangle.

12. What do you think of their statements?

Three figures have been drawn in the coordinate system below. Use a new copy of **Student Activity Sheet 4** for problems **13–15.**

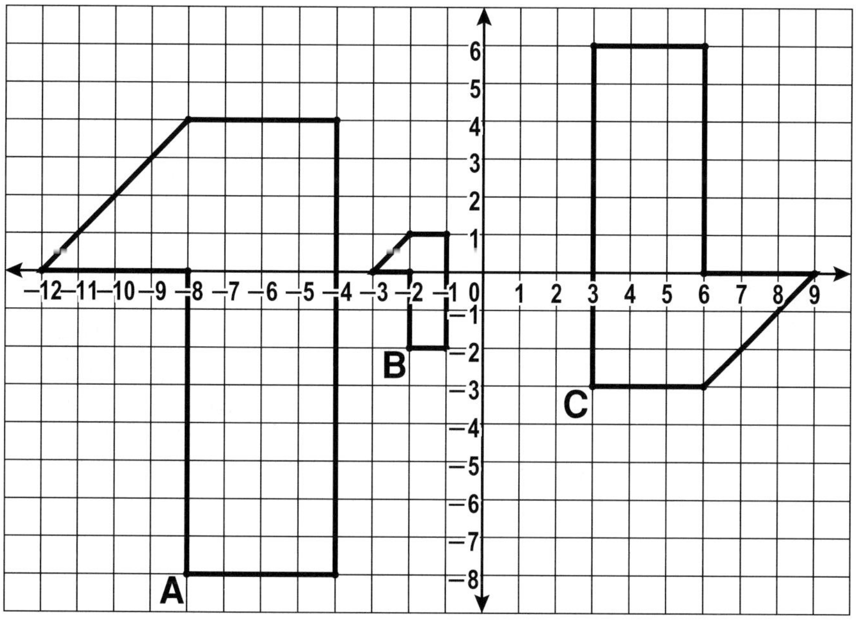

13. a. Start with figure A, B, or C. Describe how you can get the other two figures from the one you chose by multiplying or dividing the coordinates of all the points of that figure.

b. Check your answer and show how you checked it.

In the two problems below, you will multiply a figure by a number. This means that you multiply the coordinates of all the points by that number.

14. Multiply figure B by −1. Draw the new figure on another copy of **Student Activity Sheet 4** and name it D.

15. Multiply figure D by −1. What do you notice? What does this tell you about -1×-1?

Summary

To locate points on a grid, you use a coordinate system. The origin is the point at the intersection of the horizontal axis and the vertical axis. Every point can be located by using two coordinates that tell what direction and how far the point is from the origin.

You can change the position of a figure drawn in a coordinate system by adding a number to or subtracting a number from each coordinate, or by multiplying or dividing the coordinates by some number. You need to remember the rules for these operations in order to do this.

Summary Questions

16. Explain how the coordinate system works.

17. Do you think the following statement is always true? Give an explanation for your answer.

"If you multiply or divide the coordinates of a figure by a number, the size will always change."

18. How can you make a figure smaller by multiplying?

19. Suppose a figure called A is multiplied by +2 and called figure B. Then figure B is multiplied by –1 and called figure C, and figure C is multiplied by $+\frac{1}{2}$ and called figure D. How can you get figure D directly from figure A?

TRY THIS!

Section A. Graphing Figures

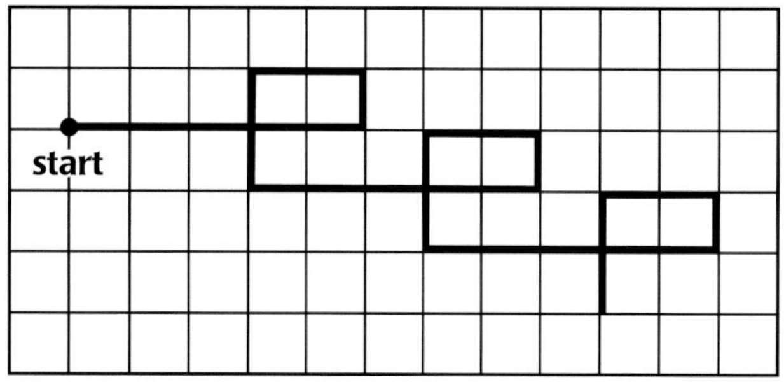

1. Write drawing instructions for the design on the right.

2. Describe the directions in which the following designs walk:

   a. +1 +3 −3 −1 +1 +3 −3 −1

   b. +2 +3 −3 −4 +4 +3 −3 −2

   c. +2 +4 −1 −2 +2 +4 −1 −2

   d. +5 +1 −2 −2 +5 +1 −2 −2

3. Create a design pattern of your own.

   a. Write drawing instructions for your design.

   b. Have someone in your home draw the design using your instructions.

Section B. Time Zones

Sharon Taylor is a salesperson for a toy manufacturing company. In order to sell her company's toys to different stores, she must travel quite often. Sharon is flying from New York to Seattle for a meeting with a retail chain called Rainy Day Toys.

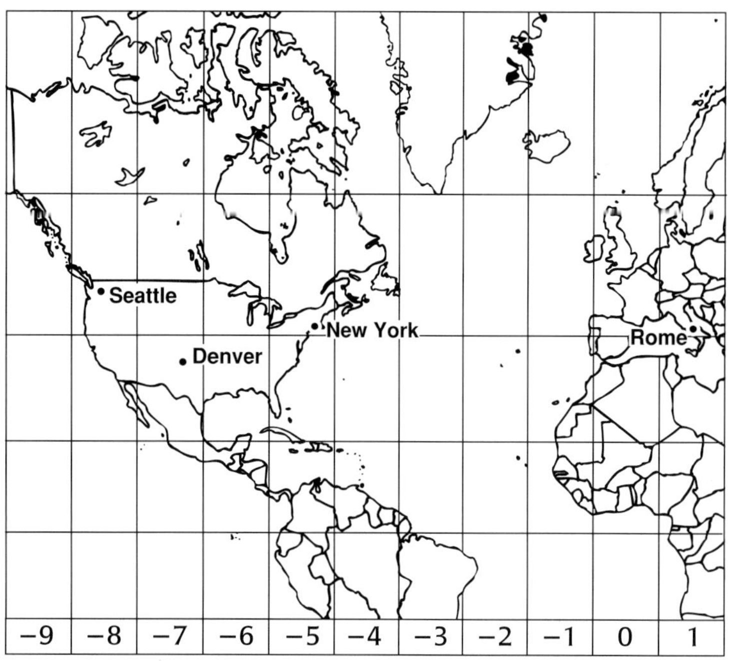

1. Sharon's plane left New York at 3:00 P.M., and it is a six-hour flight from New York to Seattle. At what time will she arrive in Seattle?

After her trip to Seattle, she returns to New York. She wants to schedule a meeting in Rome with an Italian toy retailer. Sharon works between 9:00 A.M. and 5:00 P.M. each day. The Italian retailers also work from 9:00 A.M. to 5:00 P.M.

2. When can Sharon schedule a meeting?

Section C. Plus and Minus Chips

Ms. Irvine divides her class into three groups to play a game. The students decide to keep score the same way Mr. Jones's class did: correct answers receive a plus chip and incorrect answers a minus chip.

1. After six rounds, team A's score is +2. How is that possible?

After playing for awhile, Maureen makes the following comment, "On an even round every team's score will be even, and during odd rounds the scores will be odd."

2. Explain whether her statement is true or not.

3. After ten rounds, team C has the chips shown on the right. What is its score?

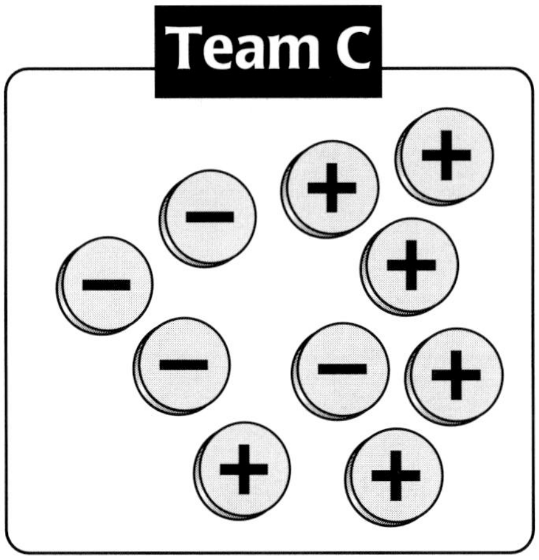

Use the team C chips shown on the right as the starting value for each part in problem **4.**

4. If the following changes are made to team C's chips, what is its score?

- **a.** Two "+" chips are added.
- **b.** Three "–" chips are added.
- **c.** Four "+" chips are taken away.
- **d.** Two "–" chips are taken away.

Below are the scores after 10 rounds for the three teams. The game has only three more rounds remaining.

Team A	0
Team B	–2
Team C	+2

5. What would have to happen in order for team B to win the game?

6. Using the scores listed above after 10 rounds, how many more rounds would it take for all of the teams to have the same score?

7. Copy the following and find two different solutions for each problem:

- **a.** +3 ___ = 0
- **b.** –2 ___ = +1
- **c.** (Positive number) + (Negative number) = –3
- **d.** (Positive number) – (Positive number) = –2

Section D. Altitudes and Temperatures

A group is planning a hiking trip from point A to point G on the trail mapped below.

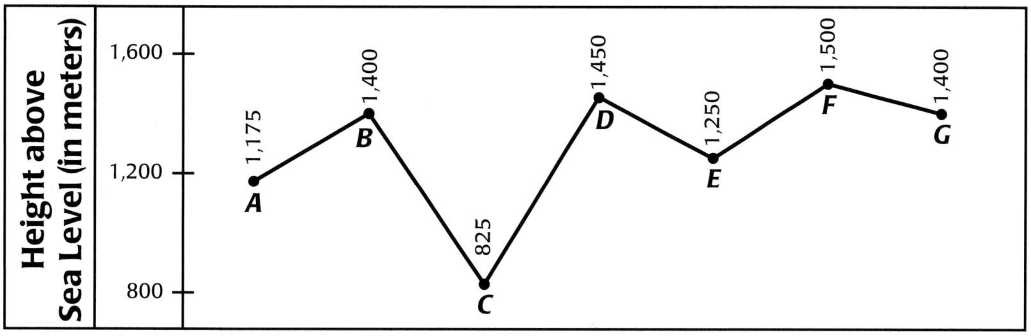

Someone in the group says, "Wow, we are going to be only a few hundred meters higher when we end. That does not sound like a difficult hike." As the hiking guide, you decide it would be helpful to explain how much climbing they will have to do.

1. Make a table that shows how many meters the hikers will have to ascend and descend. What are the totals?

Below are the elevation gains and losses for two other trails.

Trail A		Trail B	
Up	**Down**	**Up**	**Down**
200 m	350 m	50 m	125 m
50 m	100 m	375 m	25 m
225 m	50 m	175 m	350 m

2. Both trails are 5 kilometers long. Which of them would be more strenuous to hike? Explain your reasoning.

The daily high temperatures for the month of February in a mountain town are shown on the right.

3. Which week was the coldest? Which week was the warmest?

4. Complete the following calculations:

a. $3 \times -3 =$ **b.** $4 \times -1 =$

c. $6 \times -2 =$ **d.** $-20 \div 10 =$

e. $-15 \div 3 =$ **f.** $-28 \div 28 =$

Daily High Temperatures for the Month of February

Sun	Mon	Tue	Wed	Thu	Fri	Sat
				−4	−1	+1
−4	−6	0	−3	−1	−5	−1
−3	−4	−5	−1	0	−5	+1
+1	+2	+2	0	−4	−6	−1
+1	+1	+1	+2			

Section E. All Operations

1. Represent the following expressions using chips. What is the final result?

a. $-2 \times +3 =$

b. $+5 \times -2 =$

c. $-4 \times -3 =$

2. Calculate the following:

a. $-7 \times -3 =$

b. $+10 \times -2 =$

c. $-8 \times +4 =$

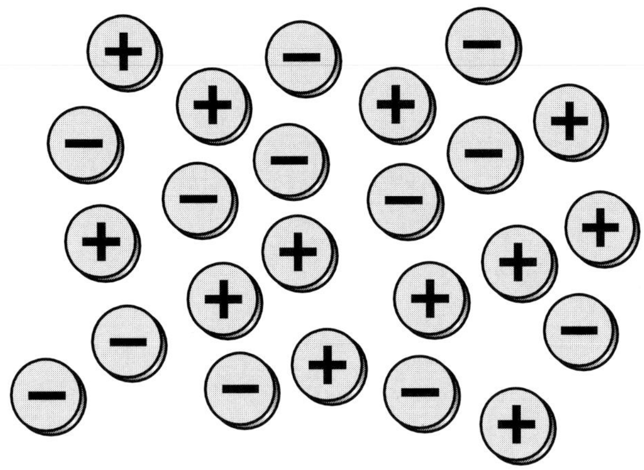

Section F. Operations and Coordinates

Use graph paper for problems **1–5.**

1. a. Draw a coordinate system. Put number scales on it and mark 0 for the origin.

b. Plot these points on your coordinate system: $A(3, -2)$, $B(-1, -2)$, $C(0, 1)$, $D(4, 1)$.

c. Connect the points in alphabetical order.

2. What are the new coordinates of your figure if you move it three spaces to the left?

3. Using the figure in problem **2,** what are the new coordinates if the figure is moved up two spaces and five spaces to the left?

Use the original polygon you drew in problem **1.**

4. a. Find a way to reduce the area of the polygon to one-half the area of the original.

b. What are the new coordinates for the reduced figure?

5. Describe what could be done to the coordinates of the figure in problem **1** to made the following changes:

a. The area is multiplied by four.

b. The figure turns upside down but stays the same size.

c. The area of the figure shrinks.

CREDITS

Cover

Design by Ralph Paquet/Encyclopædia Britannica Educational Corporation.

Collage by Koorosh Jamalpur/KJ Graphics.

Title Page

Illustration by Phil Geib/Encyclopædia Britannica Educational Corporation.

Illustrations

6, 8–9, 11–13, 15–16, 18–23, 27–34, 37–39, 41–43, 45, 47–49 Phil Geib/Encyclopædia Britannica Educational Corporation.

Photographs

1, 35 (top and bottom), 37 © These images were obtained from the Corel Professional Photos CD-ROM™; **35 (middle)** © Encyclopædia Britannica Educational Corporation.